Solar Storms:

2000 Years of

Human Calamity

by

Sten Odenwald

National Institute of Aerospace

and

The Astronomy Café

Praise for Sten Odenwald

"With his first book, *The Astronomy Café*, Sten Odenwald demonstrated that he belongs at the interface between the cosmic frontier and the public inquiry of that frontier. With *The 23ʳᵈ Cycle: Learning to live with a stormy star*, he brings us an expose on the ups and downs of the sun, our home star. With its 11-year pattern of gurgling and churning gases, we learn how turbulent the sun is, and how turbulent it can get. But most important, we learn whether or not we should worry about it. – Neil De Grasse Tyson (American Museum of Natural History).

"Odenwald not only sets a comfortable conversational tone; he adds a sense of humanity that some science books tend to omit" - Sky and Telescope

"Odenwald's concise writing and his sense of humor and layperson's writing style is a laudable public service" – Bloomsbury Review

"Odenwald offers a cogent warning, which deserves to have an impact beyond the book's own immediate readership of space science enthusiasts." - Publisher's Weekly

"Sten Odenwald has done an extraordinary job of collecting a massive amount of information from a large variety of sources. His effort will be appreciated by all who are interested in the subject of space weather." - Scientific Committee on Solar-Terrestrial Physics.

Other books by this author:

Ask the Astronomer
Exploring Quantum Space

Patterns in the Void
The 23rd Cycle
The Astronomy Café
Back to the Astronomy Café
Stepping Through the Stargate

For related information about the impacts of
solar storms on a variety of technological systems
visit the *Space Weather* website at
http://www.astronomycafe.net

Acknowledgments

I would like to thank Dr. Jim Greene (NASA Headquarters) and Dr. Bill Taylor (Raytheon ITSS) for encouraging my interest in the history of space weather. This led to some very exciting research and a long list of public talks and articles on a very 'hot' topic.

I would like to thank Dr. Joe Allen (Scientific Committee on Solar-Terrestrial Physics) for his encyclopedic knowledge of how space weather affects satellites and other common technology and helping me to understand the complexity of this problem.

I would like to thank Dr. William Murtagh (NOAA Space Weather Prediction Center), Dr. John Kappenman (Metatech Corporation), Dr. Daniel Baker (Laboratory for Atmospheric and Space Physics), and Dr. Karel Schrijver (Lockheed Martin Solar and Astrophysics Lab) for many conversations and valuable insights to this fascinating story.

I would also like to thank the Staff at the Library of Congress, Newspaper and Periodical Reading Room in Washington, D.C. for access to the newspaper resources used in tracking down the many events that were chronicled by reporters and eyewitnesses during the 1800 and 1900s.

Table of Contents

ELECTRIC PHENOMENA
IN PARTS OF EUROPE.

Telephone and Street Car Services Suspended in Switzerland for Half an Hour on Saturday.

LONDON, Nov. 1.—Scientists attribute the magnetic disturbance of yesterday to sunspots. The worst effects of the phenomena appear to have been experienced in France. Berlin was not affected, and apparently neither Austria, Italy, nor Denmark suffered.

In Switzerland, however, there occurred a most strange phenomena. The telephone service ceased suddenly and remained suspended for half an hour, while the telegraphs were rendered unintelligible and useless. In Geneva all the electrical street cars were brought to a sudden standstill, and the unexpected cessation of the electric current caused consternation at the generating works, where all efforts to discover the cause were fruitless.

Prolog

In my previous book, ***The 23ʳᵈ Cycle: Learning to live with a stormy star***, I described in some detail how solar storms and 'space weather' can cause many different technological problems during the 23ʳᵈ sunspot cycle that began in 1996. They have been responsible for electrical power blackouts, satellite failures, and radio communications problems since the dawn of these technologies.

Like so many other natural phenomena, solar storms and the beautiful aurora that often accompany them have been known to us for a long time. The most severe ones have never failed to leave in their wakes a variety of problems for humans to clean up afterwards. And as with many of the other natural disasters, they first began as annoyances, only to become severe problems as population pressure and technology relentlessly placed us into closer conjunction with them.

This is the story of some of the worst of these episodes in the last 150 years and what lies in store for us in the near future. They are not always the most severe storms known to scientists, but they are the storms that had the biggest human consequences. They are not always a single storm, but their individual consequences are often additive. While past storms and their impacts are a matter of historical record, future storms and their impacts are a matter of certainty.

The first part of this book deals with aurora and solar storms and how humans have regarded them prior to the 1800s. It is a time when little was known about aurora or the sun, and so the accounts are mostly folklore and speculation. Things change dramatically at the start of the 1800s when science had uncovered the workings of electricity and magnetism and began to carefully study sunspots and aurora. The advent of the telegraph dramatically changed how 'solar storms' made their impact upon the human day-to-day world.

The most spectacular event was the 1859 'superstorm'. In a series of accounts based upon the collected newspaper reports by Elias Loomis, I tell the story of how this event made its dramatic impact on people's lives. I also examine the diary entries of Simon Newcomb, Mary Lincoln, Amelia Ryerse Harris and Orville Hickman Browning and weave a story about how their lives also intersected with this amazing celestial phenomenon. Subsequent chapters in the book use newspaper articles to examine the many ways in which solar storms made themselves felt over the course of 150 years of scientific and technological advancement. These accounts are a rich source of information about how each storm affected the electric power grid, short-wave radio, computers and satellites, and captivated the general public. We will see that in many cases the solution to the interference was simply to develop newer technologies that were less sensitive and vulnerable to the many solar and geophysical influences. There was actually very little one could do about permanently fixing the problem in any one technology. Caught between a solar 'rock' and a terrestrial 'hard place', our only certain reaction to solar storms is to get out of their way. We can no longer plead ignorance of their consequences or their inevitability.

1   The Stuff of Legends

For thousands of years, the aurora borealis or "northern lights" have lit the skies and human imagination with ghostly incandescence. Most often seen in Arctic regions but occasionally visible as far south as the Mediterranean, they have inspired awe, fright and a fair measure of misunderstanding. We have, for example, the legendary story of how Tiberius Caesar mistook their red glows for a fire in progress and dispatched an army to Ostia in 34 A.D. to inspect damage. As recently as the 20th Century, fire departments were still being asked to chase after aurora in the sky!

The fact is, until the 1800's, not a single aurora sighted before that time actually caused any physical effects that could remotely be considered worrisome. Nevertheless, the many old accounts of how people regarded aurora before the Age of Science, is very educational reading. We catch a glimpse of how well the many observers could accurately describe the details they were seeing. Here, for example, is what an actual aurora looks like.

Aurora September 27, 1997 from Fairbanks, Alaska (Courtesy: Jan Curtis)

There are many other types of aurora as well, and the colors come from a pallet containing a broad range of reds, yellows and blues. Deep crimson auroras are the most striking and troublesome for interpretation by our ancestors, and even by contemporary observers.

Early observers and chroniclers of aurora had great difficulty coming up with words to describe what they were seeing, especially since aurora were not very common in many parts of Europe. When they were seen, it was usually through the prism of fear, which flavored many descriptions in terms of portends and bad omens. Once you have seen your first blood-red aurora covering the entire sky, it is hard to think of anything else other than your impending doom!

For instance, Gregory of Tours (538-594) was involved in clerical work and is best known for his chronicle 'The History of the Franks'. In it can be found several descriptions of what he called signs from heaven, which was the common moniker for any event seen in the sky. In his entry for 586 AD, he writes,

> *"While I was staying in Carignan, I twice during the night saw portents in the sky. These were rays of light towards the north, shining so brightly that I had never seen anything like them before. The clouds were blood-red on both sides, to the east and to the west...This extraordinary phenomenon filled me with foreboding, for it was clear that some disaster was about to be sent from heaven."* [Thorpe, 1974]

On January 1, 745 in Britain we read about "*Fiery strokes were beheld in the air, such as no men of that generation had ever seen before and were visible throughout almost all the night of the first of January.*" Britton's comments: "Simeon of Durham appears to be the first native writer to mention this aurora. Tighernac, writing in Ireland a century later, is evidently alluding to it, as he has in 745 'In the night a marvelous and dreadful sign was seen in the stars.' Some later English and Irish writers date it 743. This event appears in Seller as 'great flashes of fire rising out of the Earth'. There seems to have been other notable aurorae about this time. Tighernac has in 746 'Dragons were seen in the heavens' and in 748 'Ships were seen in the air with their men'. In Fritz' list of dates of notable auroral activity all the years 740 to 745 are included. (Washington, D.C., Mon. Weath. Rev., 56, 1928, p.403." [Silverman, 2014]

In Britain in 793 "*Very remarkable aurorae were seen in this year as the following accounts will shew:--In this year dire forewarnings came over the land of the Northumbrians, and miserably terrified the people: these were excessive whirlwinds and lightnings, and fiery dragons were seen flying in the air. A great famine soon followed these tokens. [citing Anglo-Saxon Chronicle] In the year 793, being the fourth year of the reign of king Ethelred, dreadful prodigies alarmed the wretched nation of the English, for terrific lightnings, and dragons in the air, and strokes of fire were seen hovering on high and shooting to and fro; which were ominous signs of the great famine ... [citing Roger of Howden] In the tenth year of king Brithric, there were seen fiery dragons flying through the air which tokens were followed by two plagues; first, a dreadful famine, and then the pagan nations coming from Norway and Denmark. [citing Roger of Howden] In this time of which I am speaking, there appeared signs in the country. First they were red, such as no man living had seen before; then spreading, they became as scarlet, and seemed near the earth. Then came great whirlwinds; then fiery flying dragons. And no one knew how to explain the storms and lightnings which men saw; some said that in their opinion they signified dear times; they did not speak a great untruth. [citing Geoffrey Gaimar]*" Britton's comment: "*Fiery dragons is a usual term in early writers for phenomena which are clearly aurorae borealis.*" [Silverman, 2014]

In Britain in 979 *"Simeon of Durham: After this, a cloud appeared at midnight throughout all England, at one time of a bloody, at another of a fiery, appearance, which afterwards changed to various hues and colors: it disappeared towards dawn."* Britton comment: *"Several early writers note this, the years varying from 978 to 979. The Annals of Waverley quote an exact date, October 28, 979, apparently from Siegbert, which evidently refers to the same appearance."* [Silverman, 2014]

Aurora with Hale-Bopp Comet, March 30, 1997 (Courtesy: Dick Hutchinson)

An aurora is described in the Norse chronicle, *The Kings Mirror* written around 1250 AD,

"The northern lights resemble a vast flame of fire viewed from a great distance. It also looks as if sharp points were shot from this flame up in the sky, these are of uneven height and inconstant motion, now one, now another darting highest; and the light appears to blaze like a living flame."

Strange as it may seem, when examining Old Norse literature, none of the mythological materials nor sagas mention the aurora at all.

Chinese and Korean records of apparent auroral activity from this period may also be found, for example in 1141 AD

"at night, a red vapor appeared on the night sky, ten two other strips of white vapor penetrating through the north pole and vicinity appeared also, sometimes they disappeared and then reappeared again."

A detailed account of the aurora on February 10, 1173 viewed from Londonderry, Ireland notes

"Four Masters: A great miracle was performed on the night of his death, viz., the dark night became bright from dusk till morning, and it appeared to the inhabitants that the adjacent parts of the globe were illuminated; and a large body of fire moved over the town and remained in the south east; and all the people rose from their beds, for they thought it was the day; and it continued so eastward along the sea". [Silverman, 2014]

This aurora was also seen in England. Gervase of Canterbury records:

"On the 4th of the ides of February there appeared sometime after midnight a wonderful sign in the sky. For a certain red color was seen in the air in the northerly regions between east and west. White rays also were traversing this redness, now slender like spears, now broad like tables, and now here and now there as if erected upwards from earth to heaven. The aforesaid white rays were like beams of the sun penetrating the thickest clouds." Britton comment: *"The Scottish annalist Fordun mentions the phenomenon but he is*

probably deriving it from the English writers. Short mentions this aurora and adds others in November and December; no authority has been found for these." [Silverman, 2014]

In January 1192 from England we hear that

"William of Newburgh: For in the month of January, in the year in which the king fell into the hands of the enemy, we beheld a terrible portent in the sky, no doubt indicative of the affliction which was coming upon us. For about the first watch of the night, the intermediate region of the air between north and east grew so red that it appeared to blaze, as it were; though there was not the slightest cloud, and the stars were brightly shining; and these, too, were so tinged with fiery redness, and streaked with white stripes, that they seemed to twinkle with a kind of blood-stained light. After this dreadful appearance had possessed the minds and eyes of the beholders with astonishment throughout all the borders of England, for nearly the space of two hours, by degrees gently vanishing, it disappeared, leaving much conjecture concerning it." Britton comments: *"Some early writers give 1193 and Holinshed cites it in 1194, but 1192 would appear to be the correct year."* [Silverman, 2014]

In European medieval chronicles such as *The Anglo-Saxon Chronicle* by Holinshed, we read in 1235 AD,

"In North England, appeared coming forth of the earth companies of armed men on horseback with spear, shield, sword and banners displayed, in sundry forms and shapes, riding in order of battle and encountering together there. The people of the country beheld them afar off with great wonder."

On November 4, 1322 in Uxbridge, England

"Matthew of Westminster: ... on the fourth day of November at the first hour of the night in the western parts beyond the city of London near the village of Uxbridge, there appeared in the air to many beholders a wonderful sign. For a certain pile of fire of the size and shape of a small boat, pallid, but of a livid color, rising up from the south and crossing the firmament with a slow and grave motion, set its course towards the north. Out of the front of this pile another very fervent fire of a red color and of greater quantity, similar in shape to the former, burst forth immediately with bright beams and great speed, flying through the air, which were seen quickly meeting against each other by many beholders. And by turns frequently approaching with collisions and engaging in fearful combat, the blows of which conflict and the sounds of the crashes were heard at a distance from the beholders ..." [Silverman, 2014]

Early drawing of the aurora, depicted as candles in the sky, on January 12, 1570. This spectacular aurora was seen over middle Europe and brought forth the following admonition: *"Wherefore, dear Christians, take such terrible portents to heart and diligently pray to God, that He will soften His punishments and bring us back into His favor, so that we may await with calm the future of our souls and salvation. Amen.* Another source notes "*A shocking prodigy which was seen from Kuttenberg in the kingdom of Bohemia and independently in other towns and places round about on the 12th of January, for four hours in the night. As it stood within the clouds of the sky in this year 1570.*" (Original print in Crawford Library, Royal Observatory, Edinburgh)

Nuremburg Aurora seen in December 28, 1560. Superstitious people, seeing such flashing lights in the northern sky, often explained them as sparks from the clashing of swords. This image was taken from a pamphlet published to commemorate this event. (Credit Zentral Bibliothek, Zürich)

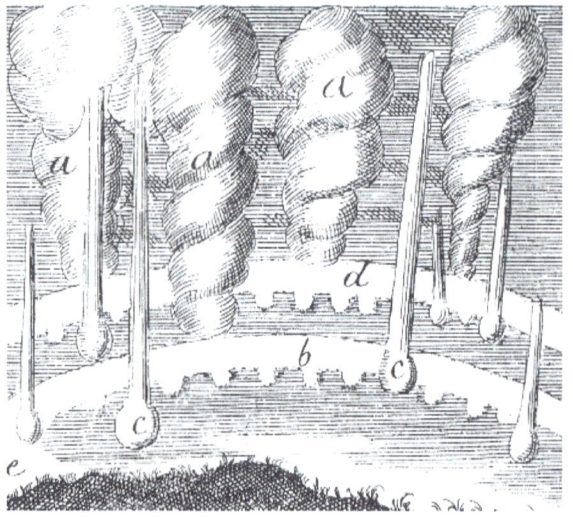

This aurora was observed on March 17, 1716 and seen over large areas of Europe. The sketch was rendered by an observer in Danzig. The many observations of this aurora have been credited with stimulating further scientific study of this phenomenon by scientists. Edmond Halley (1656-1743), reported on the aurora of March 16, 1716 and also on a second one visible from London on November 10, 1719 in the journal Philosophical Transactions of the Royal Society.

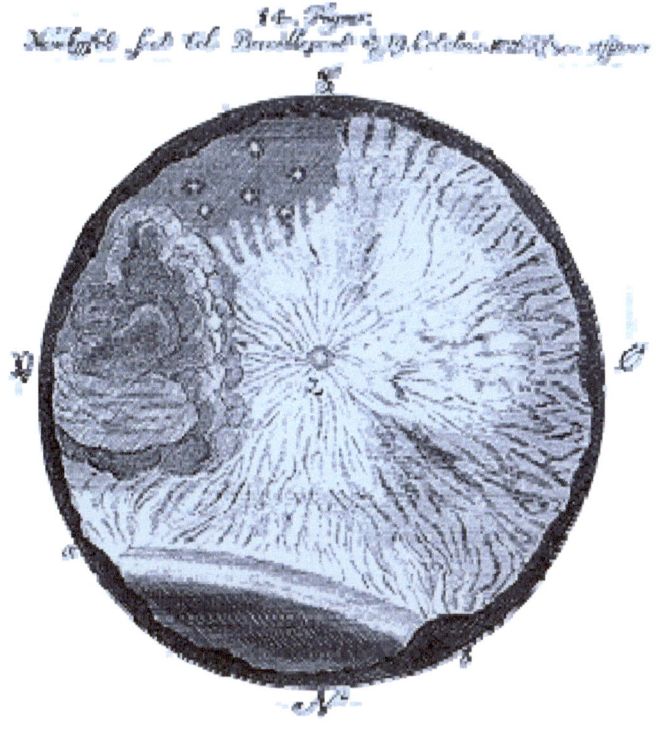

This is a sketch of the auroral corona observed during the aurora of October 19, 1726 around 8 o'clock in the evening. The crown is an optical illusion caused by many vertical streamers around the zenith location.

By the 18th century, many scientists bent their eyes towards the sky and were careful chroniclers of auroral shapes and color. In addition to Sir Isaac Newton (1643-1727), Naturalist Gilbert White (1720-1793) from rural England in his 'Natural History of Shelborne', recorded the aurora of October 25, 1769, January 18, 1770, February 15, 1779 and October 13, 1787. The Aurora Borealis or Northern Lights had previously been named in 1621 by the French scientist, Pierre Gassendi (1592-1655). They were named Aurora for the goddess of the dawn according to the Romans (known as Eos and usually described as "rosy-fingered" by the Greeks) and for the god of the north wind, which in Latin is Boreas.

Meanwhile, observations of the sun had led to the recognition that it was not a featureless disk of light, and a new force of nature, magnetism, was also added to the familiar force of gravity and was quickly made use of for practical purposes. In 800 BC we have the first record of naked eye sunspots made by Chinese astronomers. By 600 BC, lodestone and magnetic forces were cited in an ancient Chinese story of a navigation device that worked in a fog and kept pointed 'south'.

Ancient Chinese compass depicted in a modern stamp from Hong Kong.

The story seems to begin in ancient China, when Emperor Hoang-ti's troops were in hot pursuit of Prince Tcheyeou in 2637 B.C. for reasons that are now lost to us. The troops eventually lost their way in a heavy fog, so the Emperor constructed a chariot upon which stood a figure that always pointed south no matter how the chariot was pointed. Nearly two millennia later, the Phoenician sage Sanconiathon wrote, '...It was the God, Ouranos, who devised Betulae, contriving stones that moved as having life...'., and even Homer about 900 BC got into the act by mentioning this wondrous technology in the Odyssey,

'...In wondrous ships instinct with mind
No helm secures their course, no pilot guides
Like man intelligent, they plough the sea
Though clouds and darkness veil th' encumbered sky
Fearless thro' darkness and thru' clouds they fly...'

During the last thousand years, the 'secret weapon' of the Vikings evolved into the familiar magnetic compass that every Boy Scout and ocean navigator relies on to see them to safe harbor. We don't need lodestone anymore.

By the 1600, William Gilbert (1544-1603), the personal physician to Queen Elizabeth, even wrote a book about how Earth is one giant magnet with distinct north and south poles. But over the course of decades and centuries, navigators discovered that these bearings don't always run true.

We are living at a time in the history of Earth when the magnetic north-south field is very nearly aligned with the axis about which Earth spins each day. Geophysicists think that the 'geomagnetic field' is generated near the hot, electrically conductive core of Earth where hundred mile wide currents of molten nickel flow along the equator. Like many rivers of water on the surface of Earth, these subterranean currents are not steady either in space or time. Over thousands of years, even near the Earth's core, things tend to slosh about a bit. If you were standing at the magnetic north pole, you would soon discover it moves a hundred yards a day, and this forces compass navigators to buy new maps every ten years or so. Map makers and sellers since the 18th century enjoyed this aspect of geophysics quite a bit, and over time actually turned a profit from it. There are other less predictable changes that occur with magnetic bearings if you have the patience to look for them.

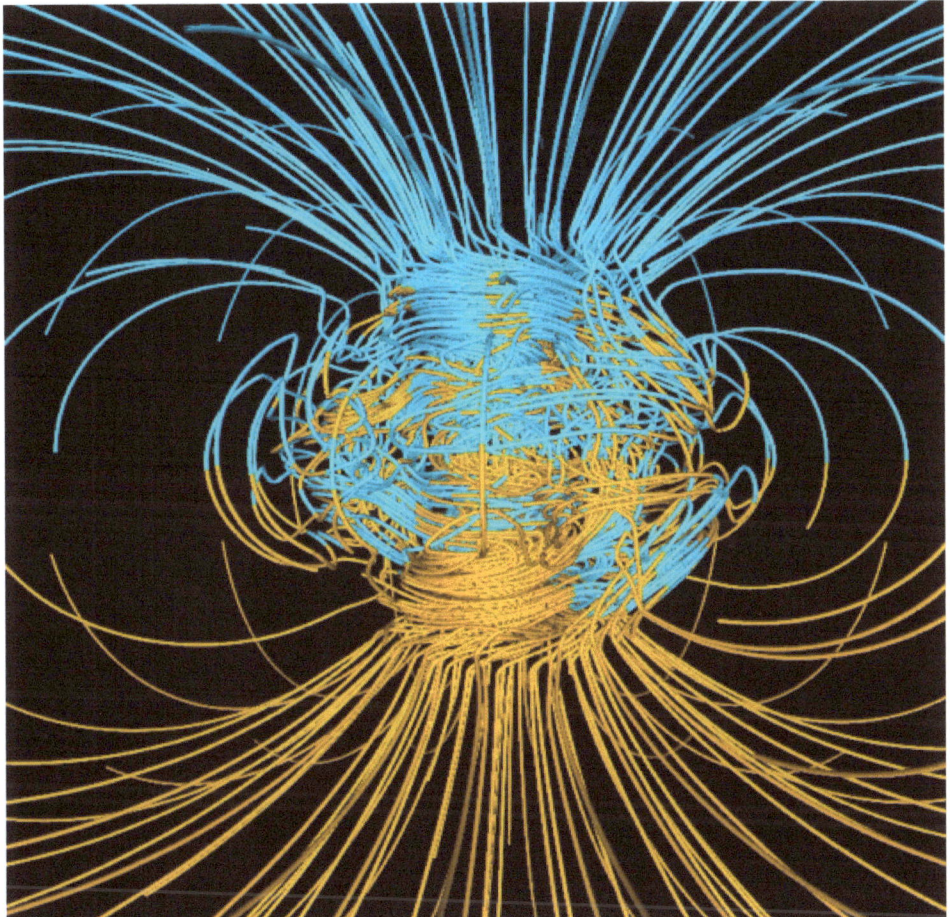

Mathematical model of Earth's magnetic field developed to study its changes over millions of years. (Credit Dr. Gary A. Glatzmaier - Los Alamos National Laboratory)

In the 1740's, George Graham (1674-1751) in London, and Anders Celsius (1701-1744) in Uppsala, Sweden began taking detailed hourly measurements of changes in Earth's magnetic declination. The fact that this quantity varied at all was known as early as 1634 by Gellibrand's observation of the 'variation of the (magnetic) variation' (Fleming, 1939). It didn't take very long before Celsius and his assistant Olof Hiorter (1696-1750) uncovered in the 6638 hourly readings, a correlation between these disturbances and local auroral activity. Moreover, comparing the records between Uppsala and London, it became quite apparent that the magnetic disturbances occurred at the same times at both locations.

In the early 19th Century, Baron Alexander von Humboldt (1769-1859), was one of those intrepid and world-renowned explorers who outfitted expeditions to Africa and elsewhere to catalog rare plants and animals. His popular stature was a combination of the measured studiousness of astronomer Carl Sagan and the down-and-dirty enthusiasm of Titanic discoverer Jim Ballard. In fact, the London Times regularly published Humboldt's weekly letters from distant lands and jungles, detailing his on-going exploits. On one of his years off from studying wild and exotic fauna and flora, his interests turned to earlier reports that compass needles didn't always point in the same direction from moment to moment. He and an assistant decided to look into this behavior a bit more.

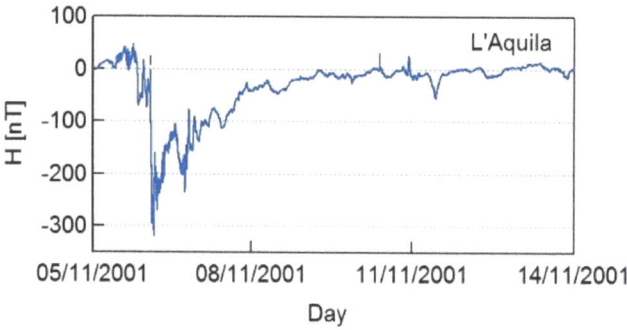

Example of a magnetic storm recorded in 2001 showing an initial sharp change followed by a several day recovery.

With a microscope, they made around the clock measurements of a compass needle's direction every half-hour for over a year. What they uncovered were the usual, and sudden, erratic swings produced by lightning storms, but every once in a while other mysterious disturbances set their needle gyrating. It didn't take long for them to realize that the strongest of these 'magnetic storms' always seemed to happen when the Northern Lights could be seen dancing outside their window or in neighboring lands to England. This behavior was taken very seriously at the time, because in terms of our monetary system today, billions of dollars of commerce were at the mercy of ships steered by magnetic compass. Within a few years, Humboldt had set up dozens of 'magnetic observatories' across the globe that were hard at work measuring compass needle gyrations and magnetic storms.

Magnetic storms are not something to trifle with. If you are a navigator, they can cause compass bearing errors as large as several degrees, so that for up to a full day, your bearings are completely unreliable and you might not even realize it. This is especially challenging and fraught with certain catastrophe if you are trying to get through a tight channel in the dark or in inclement weather. The most dramatic impact of geomagnetic storms would be a shipwreck or a plane crashing into a mountainside. Few recorded instances of such tragic events are known, however, there are stories about a ship that ran aground on Bear Island just before World War II, and airplane pilots in Alaska have claimed that some crashes were caused by just such geomagnetic storms. The problem is that historical accounts of geomagnetically-induced navigation problems are almost entirely anecdotal. The earliest account is reported in the American Journal of Science and Arts by a contributor named, simply, 'A. de la Riva',

> *"M. de Tessan cites an observation made in 1818 by M. Baral, another French naval officer, on the same coasts of New Holland, who found that he had been making a wrong course from following his needle...But on the evening of the same day, there was a brilliant aurora, and to this he attributes the deviation,"*

So, were it not for that supremely useful technology called the compass, humans would have no clue that aurora could have anything more than a psychological effect on human behavior. During the 1800s, however, a new technology would become indispensable to our needs and make an even more compelling connection to the invisible forces operating around us.

Because aurora, and the powerful electromagnetic forces that cause them, have a strong affinity for all things electrical, it is not surprising in retrospect, that every communications technology we have deployed in the last 160 years has fallen victim to interference from these natural events.

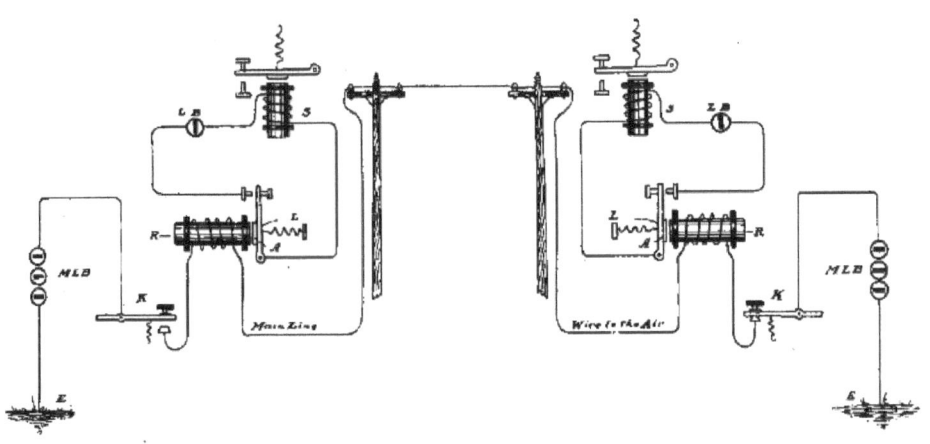

The Main Line Circuit.

This simple circuit was the backbone of global telegraph technology for decades. Although the main single wire was strung between telegraph poles to form one-half of the electrical circuit, the other half of the circuit was 'grounded' to the earth. An earth current flowed underground between the stations to complete the circuit. The problem was that solar storms also produced earth currents, and these entered the telegraph circuit to cause all sorts of problems!

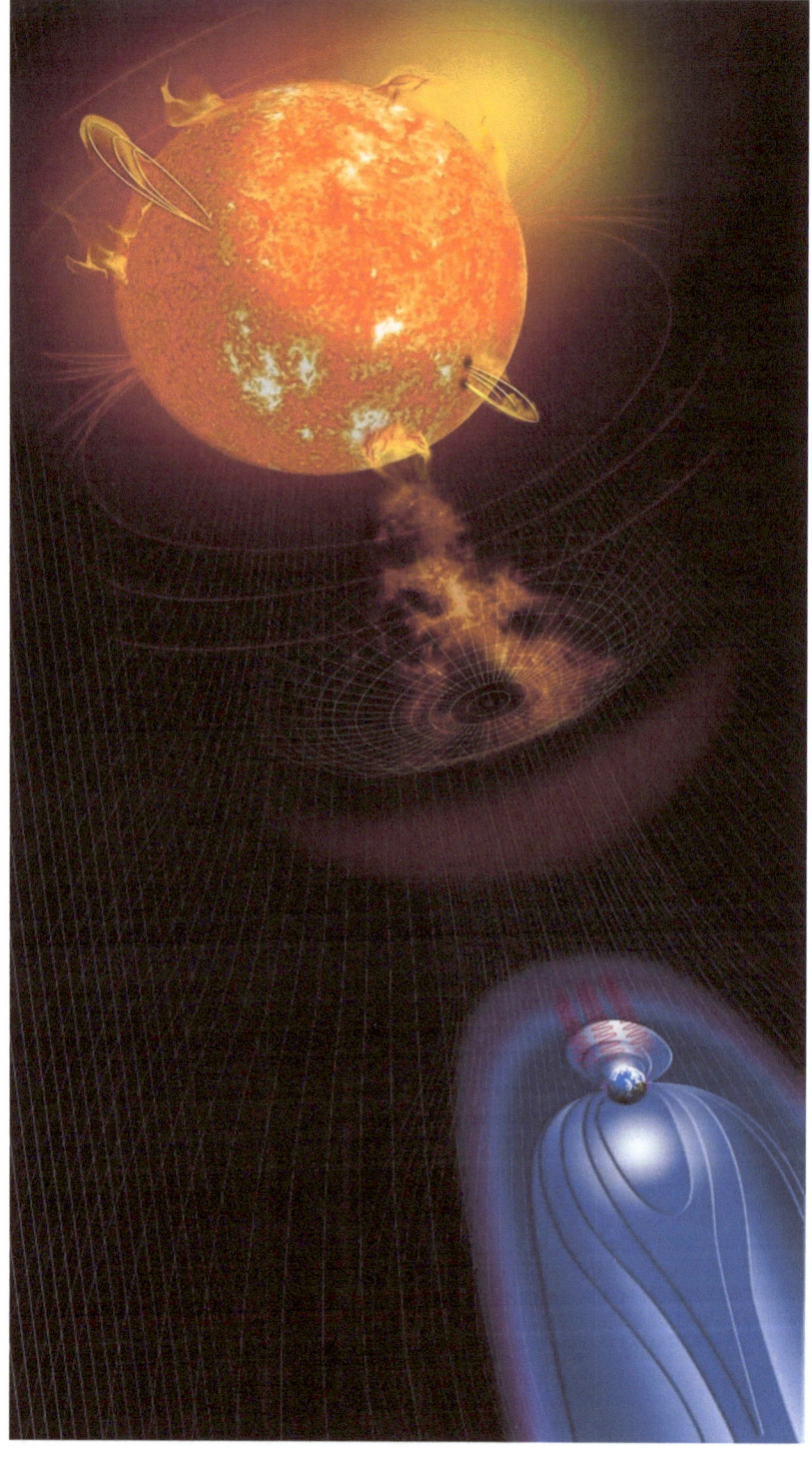

Stylized view of solar storms and magnetic effects. (Credit: NASA)

2 The Modern Era

Electrical currents were magical in many of the ways that they worked. For example, in 1820, Hans Oersted (1777-1851) a physicist at the University of Copenhagen could make electrical currents deflect compass needles. Meanwhile, across the English Channel, Michael Faraday (1791-1867) uncovered an equally mysterious electrical phenomenon: If you move a magnet across a wire, it causes a current to flow in the wire. It's hard to imagine the excitement these investigators must have felt as they saw electrical currents produce invisible magnetic forces and vice versa. Faraday's discovery of changing magnetic fields producing electrical currents, combined with Alexander von Humboldt's discovery that sudden changes in the Earth's magnetism can occur in 'magnetic storms', provided the ingredients for an interesting natural experiment. All that was needed was a network of wires large enough to catch nature in the act of inducing currents. The 30,000 mile long telegraph network available in 1848 provided just the right technology for the experiment, and during the next few years, telegraphists caught much more that simply the dits and dahs they had bargained for. For a long time they had no clue what was going on in their wires.

The Lines of Morse's Electro-Magnetic Telegraph - From The Weekly Herald, New York, January 29, 1848.

An Aurora Borealis observed at Bossekop (Finmark, in northern Norway), on 19 January 1839.
From 'Electricity and Magnetism' by Amedee Guillemin (1826-1893), published in London in 1891.

During the aurora of November 17, 1848, the clicker of the telegraph connecting Florence and Pisa remained stuck together as though it had become magnetized, even though the receiving apparatus was not in action at the time. This could only happen if an electric current from some outside source had flowed through the wires to energize the electromagnet. Telegraphers elsewhere also began to notice that their lines mysteriously picked up large voltages that caused their equipment to chatter as well, with no signal being sent. Much of this was soon attributed to the long wires picking up lightning discharges in their vicinity, and the solution was simply to erect lighting rods on the telegraph poles.

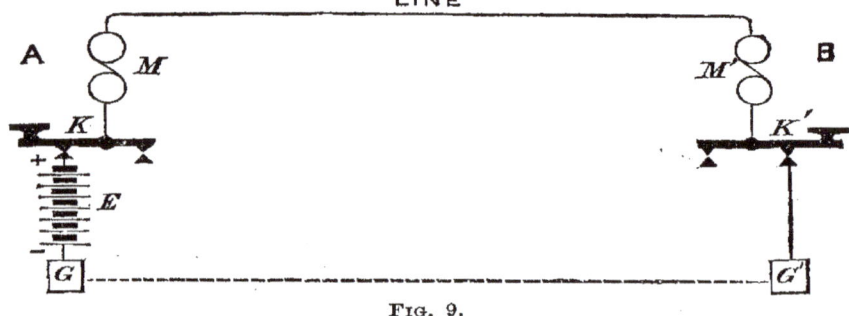

FIG. 9.

This diagram shows two telegraphs connected by an overhead line. The dashed line shows how the battery-operated (E) electrical circuit completes itself by using the earth's natural ground current G to G'. The problem is that during an aurora, earths ground current increases due to the magnetic storm effects.

Meanwhile, in 1851 Heinrich Schwabe (1789-1875) used the accumulated records of sunspots to uncover the 11-year sunspot cycle. This mysterious cycle was the first new thing learned about the sun in thousands of years. Then in 1856 Edward Sabine (1788-1883) combined observations of aurora sightings with sunspot data and discovered that aurora were correlated with the sunspot cycle. More aurora can be seen when sunspots are numerous, than when there are few spots. The figure below shows a modern plot of the number of sunspots in yellow, with the number of aurora plotted in red. The two data are almost in step with each other, but the peak of the aurora curve is slightly later than the peak of the sunspot cycle by up to one year or so.

ANNUAL SUNSPOT NUMBER & Ap DAYS >= 40

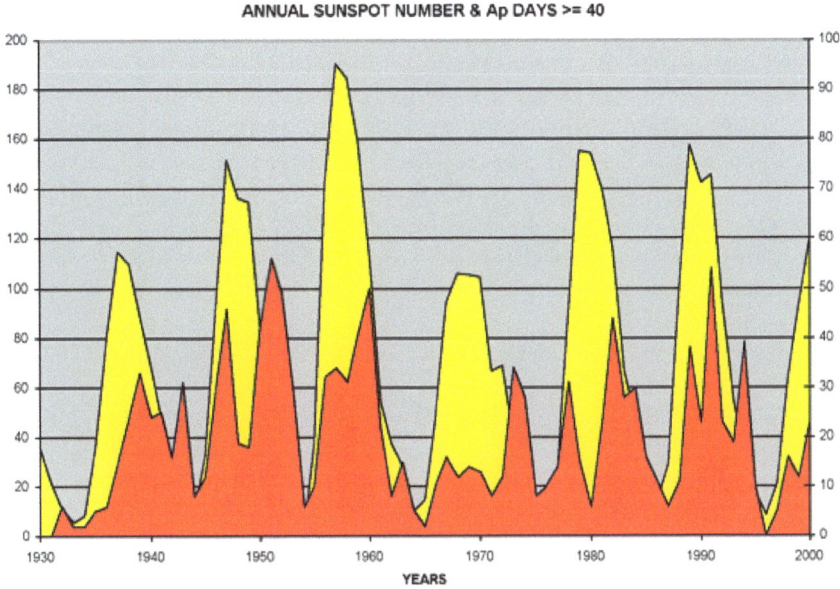

The first known forecast of earth current severity based on solar activity was offered in 1879 by William Ellis of the Royal Greenwich Observatory. In a short article written in the Journal of the Society of Telegraphic Engineers and Electricians (vol 8, p. 214), he informed the telegraphic community that sunspots are correlated with periods of strong auroral activity, and that the next sunspot cycle was coming to a maximum of activity in 1882. He noted that in the most recent years, there was little magnetic activity, and that telegraphic technology had taken a turn towards even more sensitive apparatus, as he noted

> "I would therefore ask whether any of the new apparatus possesses such peculiarity in their principle or construction as would render it more liable than were the older forms to be temporarily deranged or interfered with by earth currents?"

American telegraphists had only a short time to puzzle over atmospheric electricity on their 1000-mile lines when in 1859, the Great Auroras of August 28 and September 4 blazed forth and lit up the skies of nearly every major city on the planet. It was one of the most remarkable displays ever seen in the United States up until that time, and its effects were simply wonderful. These aurora were so exceptional that the American Journal of Science and Arts published no fewer than 158 accounts from around the world describing what the display looked like, the telegraphic disruptions they produced, and assorted theoretical speculations. Normal business transactions requiring telegraphic exchanges were completely shut down in the major world capitals. In France, telegraphic connections were disrupted as sparks literally flew from the long transmission lines.

It was generally accepted that only lightning storms could severely cripple a telegraph line. Joseph Henry, Secretary of the Smithsonian Institution, had been asked to look into this problem. His paper published a decade earlier, described the details of how thunderstorm currents infiltrated the wires, and how lightning rods could be used to reduce this effect. This expensive solution seemed to work at first, but no matter what was tried, there always seemed to be a persistent electrical' noise' on the wire that was not so easily evicted. Charles Chester later commented in a scientific journal article that, *"Slight variation of current exists almost always, generally less at night than during the day, and being more prevalent on long than short circuits, they may be attributed to the electricity of the atmosphere"* Technology was being rapidly developed to deal with this problem in due time, but there had been hints that certain other natural phenomena could, from time to time, wreak havoc with the network. Not long after Henry's suggestions became public, during the Great Aurora of November 17, 1848, Carlo Matteucci (1811 - 1868) the Director of Telegraphs in Pisa, observed a new form of disturbance.

Matteucci was by that time a world-famous authority on electricity and the workings of the 'electric pile' - the forerunner of the battery. Through a recommendation by Alexander von Humboldt to the Grand-Duke of Tuscany, he was offered the Chair of Physics at the University of Pisa. While continuing his researches into electricity, he began making breakthrough discoveries in electro-physiology, showing that Galvani's animal electricity was real, and that electricity accompanies the heartbeat. What he accidently discovered with telegraphs during the 1848 Aurora, however, was both mystifying and intriguing.

A telegraph set contains a metal lever that is pushed by the finger so that its tip makes contact with a metal button. This closes the circuit with the telegraph line so that a current is sent to a remote station. There, it causes a coil of wire in a 'sounder' to be activated as an electromagnet. This causes a corresponding metal bar, called the armature, to make contact, creating the familiar 'click'. What Matteucci discovered during the aurora was that the armatures of the telegraph instrument connecting Florence and Pisa remained stuck together as though they had become magnetized, even though the receiving apparatus was not in action at the time, and the batteries were disconnected. This could only happen if an electric current were flowing through the wires to energize the coil of the electromagnet. It would have to be a strong current unlike a bolt of lightning, but far stronger than any known battery voltage. The fact that this happened at the same time as the aurora, which glowed ominously in the Italian skies did not go unnoticed by Matteucci, though little mention was made of this 'coincidence' elsewhere. In the years shortly before the August 1859 aurora, the evidence for a direct link between aurora and telegraph disruptions was still considered sparse and inconclusive. Even the experiment performed by another famous French physicist, Auguste de la Rive in 1854 did not produce much of a stir.

In the early-1800's, Auguste De la Rive (1801-1873) was one of the founders of the electrochemical theory of batteries. He experimented with the voltaic cell and supported the idea of Michael Faraday that the electricity was the result of chemical reactions in the cell. He invented a prize-winning electroplating method to apply gold onto brass and silver. He determined the specific heat of various gases, examined the temperature of the Earth's crust, and made ozone from electrical discharge through oxygen gas. He was a contemporary of Faraday, Ampere and Oersted, with whom he exchanged correspondence on electricity.

Perhaps in an effort to explore the mysterious world of 'Earth currents', De la Rive had decided to attach an instrument for measuring electrical current to telegraph cables in England. What he discovered was that, during auroral displays, not only do magnetic storms occur, which had been commonly studied by scientists by this time, but 'auroral currents' can be made to flow in telegraph lines. *"But the most remarkable fact, is the perfect concordance which these observations have proved to exist between the movement of the needle of the galvanometer placed in the circuit of the telegraph wire, and the diurnal variations of the magnetic needle [produced by aurora"* But no one had ever imagined that even more severe disruptions could occur, because no one really knew just how severe an auroral 'storm' could get. Besides, in 1854, there were 27,000 miles of telegraph lines in operation in the United States, with fewer than a million messages crossing them each year. Although its usefulness was growing at a fantastic pace, at the moment telegraphy was hardly a technology in high demand, so any impacts to it by aurora were limited to only a small community of users who generated a few scattered anecdotes.

Meanwhile, the first photographic technique was developed in the 1830's by J. N. Niepce (1765-1833) and Louis Daguerre (1789-1851), and relied on the exposure of a thin iodine layer deposited on a silver substrate, subsequently fixed in a mercury bath. The images so produced became known as daguerreotypes. This imaging technique was very soon applied to astronomy, through the enthusiastic support of the French astronomer and politician Francois Arago (1786-1853), and the British astronomer John Herschel [(1792-1871), son of William Herschel], who first coined the term "photography." The first

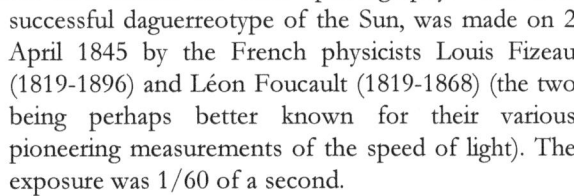

successful daguerreotype of the Sun, was made on 2 April 1845 by the French physicists Louis Fizeau (1819-1896) and Léon Foucault (1819-1868) (the two being perhaps better known for their various pioneering measurements of the speed of light). The exposure was 1/60 of a second.

The original version of this image shows the umbra/penumbra structure of sunspots, as well as limb darkening.

Martin Brendel is credited with having taken the first successful aurora photograph on February 1, 1892. This work was followed by Carl Stormer's campaign to create over 1000 photos of the aurora for detailed scientific study. The first successful color photographs didn't arrive until the 1951 work of Gartlein and Petrie who presented their color photos at Auroral Conference in London, Canada. Life magazine published the first color photos in the June 1953 issue of the magazine. Since then, amateur and professional photographers have taken millions of photos of the aurora and created spectacular collections of their work on the World Wide Web for all to see and appreciate such as this one taken by Senior Airman Joshua Strang; a member of the US Air Force in Alaska.

What I would now like to present is the interesting lead-in to one of the greatest solar storms ever recorded in the last 200 years. Thanks to many diaries available at the Library of Congress Manuscript Division, it was possible to find several active diarists around the time of this great solar storm event, and attempt to paint a human picture of how different people reacted to this event beyond the many inevitable newsprint interviews.

3   The 1859 Superstorm

August 28, 1859 When he woke up this morning he had no idea that by day's end he might be electrocuted; another life snuffed out by circumstances not of his own making. For the moment, in the reverie of morning's dawn, he knew only that he was Frederick Royce: a telegraph operator in Washington DC soon to be working the dreaded 12-hour late shift. At 18 years of age, he was living alone in the city's mosquito-infested suburbs on an otherwise typical, late-summer's day. The Capitol was still under construction, and there was hardly a place to walk without getting filthy from mud and other debris.

Photograph of Washington DC Capitol construction ca 1860. (Courtesy National Archives Collection)

Telegraphy was an uncommon profession in an uncommon town. He had left his family home in Pennsylvania to prepare for this elite profession; an arduous and queer training to learn Morse's Code. Along with thousands of other teenaged boys across the country, he had hung out at the local telegraph office watching the Operator key-in the enigmatic messages, and listen to the dits and dahs sound-out on his instruments. It was a captivating experience. Like a locomotive engineer or a steamboat pilot, Telegraph Operators were an immediate attraction to boys. The boys, and much later a growing cadre of girls, had all sensed that this was The Future paying them an advanced visit. They built their own clickers, and entertained themselves by learning the Morse Code on their own, which confounded their elders and often got them in a heap of trouble in school. A few experiences with the lash soon convinced them that teachers did not appreciate students passing notes in a secret code.

Some young boys were fortunate enough to have a real Operator as a mentor, but for many, they attended special schools and classes to learn what they needed to know. That's how Frederick learned his skill; at a special big-city school in Washington DC. After a grueling pace of memorization and constant drilling, he had hit five words per minute after 30 hours of study, and 15 words per minute after three months. The hardest part wasn't in the sending of the message, but in the receiving of it. To the untrained ear, the dits and dahs ran together into a jumble of clicks barely separable in tempo. By far it was the hardest part of the training. He knew from his mentors who had been doing this for five years that some Operators could hit 25 words per minute. But at his pace he had qualified for employment and that was all that was needed for right now. For this effort, he became one of the youngest operators at the House Office building.

As his 'plug' status would have it, he drew the worst shift. There were frightfully many messages to send out from the always-busy Washington office, and not enough lines out of the station to really accommodate the traffic. Telegraph technology was only 15 years old, and already some five million messages flowed across the US network every year. For the next twelve hours, Frederick would be in charge of his little corner of this vast exchange of letters, messages, government notices and whatever else his largely unappreciative clients felt like sending. He would often muse to himself as another rude client asked for the impossible, how quickly we have become accustomed to such a new technological miracle!

He looked at the message he had just been handed with a destination of New York. It was a simple little note written in the compressed short hand that they had all learned with most of the vowels dropped, and certain codes inserted for the common, longer words. Without thinking, he grabbed the plug for his key set and inserted it into the receptacle marked 'New York' on the patch board in front of his desk. At this time, he wasn't even thinking of the English letters scrawled across the dispatching slip. All he experienced were the mental twitches that he would shortly be using to key the message onto the line. Morse's Code had become a second language to him in a way that few people could comprehend. One after another, he keyed the dits and dahs of the client's message into the table top mechanism. A tidy profit of 40 cents was racked up for the Company, and another 4 cents added to his salary.

Everything seemed to be going well until around 7:30 PM when the lines started acting up in a weird way. He had never seen anything like this before in his rather brief career. Each time he released the clicker, which cut off the battery from the Richmond line, a spark would jump from the contacts as though the line had been connected to a powerful battery far greater than the ones down in the Battery Room. He asked the Night Chief Operator what might be going on, but in reply he only mumbled 'There must be a lightning storm going on somewhere, and the electric fluid has gotten into the wire.' It sounded like a perfectly good explanation to him. He didn't really understand how electric telegraphs worked anyway – no one really did. That was a job better left to the inventors and the scientists, he thought. So he just kept on doing what he was paid to do to the best of his ability, and under ordinary circumstances he was pretty good at what he did. But when the line is full of interference the way it was right now, the clicker clicks the dits and dah's and lots of other chatter, and fills the message with nonsense letters.

What happened an hour later at 8:30 PM was not something he ever imagined in his wildest dreams. There were many kinds of accidents that could befall the unlucky in this day and age. All you needed to do was read the newspapers. Farm equipment ran over farmers, steam boilers exploded, children were decapitated while playing on railroad tracks. Nature was also inventive and took its toll on the unsuspecting and incredibly frail human form. But injury or death from electricity other than from a lightning strike was completely unfamiliar. He knew that the telegraph worked by electricity. You could go downstairs into the smelly Battery Room and see the electricity being created from bubbling chemicals in the hundreds of noxious Grove cells. But at the Operator's station, you could touch the bare copper lines and not feel a thing. It was not in anyone's common experience in this era of steam power and horses, to be jolted by an electrical current.

So, when Frederick was working hard to get the next message out against the interference, a discharge of electricity suddenly leaped out of the ground wire. It struck him squarely in the head, leaving a pea-sized patch of singed skin in its wake. The current had entered his head and made its way inside his body to the ground. It was an unexpected, and powerful jolt that left him dazed though miraculously not electrocuted. The old man who was his client and who was sitting nearby, watched the whole thing in astonishment, but didn't even give him a tip for his effort. Did the customer think that this kind of thing happened to telegraph operators all the time?

Elsewhere in a different corner of space and time, a poet stared up at the night sky in disbelief, groping for words to describe the kaleidoscopic visions that assaulted his senses. The dazzling procession of colors and shapes continued unabated for hour upon hour: Streamers and undulating celestial arches, glowing crowns of light and shooting meteors. From sunset to sunrise, the choreography of light continued in perfect silence, with a brilliance that at time lit up the entire landscape in a terrifying blood-red color.

William Ross Wallace was more in touch with his feelings and sensibilities than most people. He was the master of the written word and turn-of-phrase, and he had honed this rare talent for a decade. From many quarters, his poems appeared in magazines and newspapers with their almost lyric and musical cadences, which was his trademark. This night, however, the heavenly dance of light would greatly tax his talents. He had to find just the right words, assembled just so, to capture the awe and sublime mystery of the silent spectacle. What would be the theme of the poem? What should his message be? What were the aurora above saying to him in a way his readers would also find meaningful? And then as the dance entered a new level of activity, it finally came to him.

Within an hour, and with the lights shining into every nook and cranny in his study, he wrote his lines and couplets. The message from above wasn't about some transcendent fear or implied doom. By daybreak, his energies had spent themselves as the last of the words flowed from his pen. The draft was scrawled almost illegibly with many revisions of words and orderings of lines in his struggle to find just the right flow, just the right meter. He made one last read of his final cleaned copy, folded it neatly into an envelope, and after affixing a 3-cent stamp, posted it to the New York Times. It was, to his mind at the moment, an important poem. For the next week he continued to dwell on the sentiments he had explored, but then like a dream that fades slowly at daybreak from consciousness, it was time for him to move on.

While the Poet basked in the aurora's powerful and sublime messages, some of the inhabitants of New York, Philadelphia, Cuba and Puerto Rico were dealing with more pragmatic fears. The inhabitants of these cities don't normally have much in common. On the one hand, wealthy New Yorkers and Philadelphians went about their normal, but complicated, city lives. On the other hand, natives of Cuba and Puerto Rico scratched out their existence as best they could, often in the filthiest state of abject poverty. But tonight they all experienced the same fear: The crimson skies to the north were vivid and rapidly changing. You could see the glow reflected in the ocean, or on the white-washed buildings. Could it be possible that their cities were being engulfed by catastrophic fires? Alarms sounded and firefighters dashed out of their stations in search of conflagrations of fire that were not really there. Like a mirage, the more they drove to meet them, the faster they seemed to recede northwards. But what if these events were, instead, portends and omens of dreadful things to come? The minds of the impressionable raced headlong to other conclusions even more terrifying than local city fires.

In Columbus Ohio, a 16 year old girl tried as hard as she could not to think of The Apocalypse and the end of the world, but no matter what verses of the Bible she attempted to substitute, the blood-red lights relentlessly brought her back to the only possible meaning behind it all. The serpentine shapes twisted and writhed like serpents from Hell. And from Revelations she could recite verse after verse that described the Seven Signs. Would the final moments be painful, she wondered? Would she be among the sin-free who would ascend and be Saved? There was no way to stop the thoughts from erupting in her mind. They would subside for a moment, then the deathly crimson light would invade the space beneath her bed, and the terrible thoughts would resume anew as if drawing some evil power from the glow itself. Finally, she could stand it no more. As the roosters crowed at the light of the crimson false-dawn, and in a desperate move to escape and warn others, she sprang from her room and ran through the streets; an uncontrollable wretch screaming hysterically in the night. The Sheriff finally cornered her, and brought her to the jail house. From there, she would be taken to the local asylum where she would spend the rest of her life in search of the solace that would never come.

It was 1859, and for a handful of nights in late-summer the world became a crucible. So many things had been mixed together, so much awe and misery. It would be a century before wiser men fully understood what events had circumscribed this world and pitched it into such chaos. But the days that led up to this mingling of unlikely events around the globe were utterly unremarkable.

For those interested in watching the sky either as serious stargazers or merely sitting by the campfire, a rare planetary alignment was holding court in the morning twilight – often a portent of interesting, or ominous, events depending on your point of view. Moments before sunrise, Venus, Mars, Saturn and Jupiter stretched like pearls on a necklace from Leo to Cancer. Venus near its maximum brilliance, and Mars a much fainter sight were closest to the sun, and discernible in the bright glow of twilight. Saturn and Jupiter farther away would have made eye-catching objects in the early morning sky, shining with their intense silvery light. A crescent moon commanded attention between

Jupiter and Saturn. In a week, the planetary lineup would gain Mercury as a new, though straggly, member, having just escaped from the evening twilight skies. Had this alignment caused much of a stir? There was not the slightest mention of it to be found among the more mundane news stories carried in the local papers.

The steamship Great Eastern was soon to embark from England enroute to either Portland or New York.- newspapers could not seem to make up their minds which it would be. It was the Titanic of its time, carrying 1000 passengers on a multi-week luxury cruise across the Atlantic.

Meanwhile at Niagara Falls, The Great Blondin would be attempting another perilous tightrope walk across the falls. Would he fall to his death, or make it across? The public followed the story with excitement as it unfolded in the newspapers. Hotels in Niagara sold out as people flocked to see his latest spectacle, and to watch with ghoulish interest hoping for the unfortunate accident that was surely to come...but never did. Blondin's latest feat was to prepare breakfast on a stove while suspended over the frothing tide of Niagara. Following his lead, many others attempted daring feats of their own as a minor epidemic of copycat stories began to emerge. Rope-walking mania swept the country, and children on backyard ropes broke their arms and skulls trying to emulate their daredevil hero.

These events were, themselves, set within the more complex background of the year 1859, which found 20 million Americans and 3 million slaves at a confluence of circumstances that would grow in consequence as time moved forward. The southern states were on the verge of declaring themselves The Confederacy. Darwin's book 'Origin of Species' was at the printing office being prepared for a First Edition November release of a mere 1250 copies; a book whose destiny was to rock the world for a century to come. And elsewhere, John Brown was planning his uprising at Harper's Ferry in October.

There were also 40,000 miles of telegraph wire connecting 1500 telegraph stations across the country, and 60,000 more miles of lines scattered across the world. As a nation we were still beholden to over five million horses to carry us from place to place, and either railroads or stagecoaches to ferry most of our mail. Meanwhile in Titusville, Pennsylvania, Edwin Drake drilled the first oil well on Earth, and on this day struck oil; a simple act that changed the history of the world.

As luck would have it, the congress of planets and the moon in the late-August morning sky made the night so deep and dark that eyes could drink in the faintest stars possible. It would be the dim glow of starlight that would illuminate the path taken by nighttime strollers who, for whatever urgent reasons, decided to navigate the streets or forest paths. There were no omens of the impending event to be seen in the mysterious movements of planets in the celestial firmament. Nothing could be found in the mundane actions of individuals in the terrestrial sphere, ether. People had better things, and certainly more practical matters, to be concerned about in the pre-electricity days leading up to the aurora. For most people able to watch the skies, life was difficult-to-hard. In many quarters, survival itself depended on a good night's sleep.

A team of prospectors lured by the early summer announcement of gold in the Colorado Rockies, made their laborious way along an unknown escarpment, leading a team of mules well-stocked with food and supplies. Horace Greeley had been this way before, looking for a story to publish in his big-city newspaper, or perhaps looking for a nugget or two himself. Many people had come here since then, searching for their own Big Strike in the shadow of Pike's Peak. The thinning air at 10,000 feet made it difficult to travel more than a few miles at a time before having to stop. Lush alpine meadows brimming with the insect life of late summer were at first a pleasure to see until the dreadful biting of mosquitoes and black flies began!

The gold seekers trudged dusty mountain trails with vague destinations in mind. Perhaps a brisk stream offered an invitation to set up a temporary camp, get out the pan, and sift the sands for a few flakes of gold. The glint of a vein of ore catches the eye of a seasoned prospector. Out comes the pick to pry out a sample. Is it gold? Is it worthless pyrite? Some miners entered the Rockies and left with thousands of dollars of gold in exchange for months of hard labor and bad food. Others disappeared into the alpine wilderness never to be seen again. It wasn't that bad things had happened to them, though local Indian tribes were not always friendly. For many trusted guides, the promise of finding riches around the next bend was a siren's song that could seldom be ignored. Stories were told of the amusing events that surrounded the Pike's Peak gold rush that had been underway since mid-June. People were dispatched from Denver City to check out the truth of the story. When they didn't return, more were sent and they, too, remained in the mountains having joined hundreds of others near Gregory's Diggings, Jackson's Diggings or Boulder Town. Denver City became a crossroads for despairing outgoing miners selling their supplies to return to civilization, and incoming miners-to-be who were willing to buy the same dog-eared and worn supplies for a few bucks to try their own luck. Some people endured the excitement and hard work of the quest for gold. Others found themselves locked into very different worlds of obsession and daily ritual.

Orville Hickman Browning had been many things in his day: Whig politician, Senator from Illinois, lawyer, and not inconsequentially, a friend of Abraham Lincoln. Orville had just returned to his home in Quincy from a six-hour trip by railroad to Springfield on August 24th. He had been asked to make the trip by State's Auditor Jessie K. DuBois. It was a last-ditch move to prevent Treasurer James Miller, from resigning. Arrangements of an entirely undisclosed variety had been made to have Miller resign so that Governor Bissell would appoint Alonzo Mack as his successor. Orville had strong feelings about this kind of swap because he didn't think that Mack was at all fit for the office. Besides damaging the state of Illinois if Mack were appointed, he would probably also damage the reputation of the Republican Party through his incompetence in such a high position of civic responsibility. A hasty, late-morning meeting with Dubois and Miller had assured Orville that Miller would not resign after all. Yet as it would turn out, this would not be the end of it. Dubois and Lincoln did not have much faith in the old man's promises. Instead, they shared the suspicion that Miller's resignation was already in Bissell's hands. In what seemed like an inevitable outcome, Mack did assume the office of Treasurer of Illinois on Thursday. At the same time, Orville and his wife Emma were taking the cross-state 6:00 AM train back to Quincy, arriving there around Noon.

Thursday, August 25 was like so many others in his hectic life as a lawyer. It found Orville at his desk in his cluttered, though elegant office. But unlike most days, he had spent a thoroughly enjoyable hour taking afternoon tea in the English tradition. How odd, it seemed, that we had fought a war of independence with the English, only to surrender to many of their finer, and civilizing, traditions. Friday was a day of hard office work, capped by an evening soirée with Miss Corey, Miss Austin and Mr. Tucker who were all visiting

Quincy from out of state. The conversations in the Browning parlor and drawing room ranged from the trivial to the consequential. It was a welcomed respite from the aggravating business of political in-fighting that had started the week in Springfield, and the vexing issues of the murder case against his client Joseph Hollingsworth in Mercer County that would soon follow.

August had also been a busy month for Amelia Ryerse Harris in London, Ontario though of an entirely different order. She wasn't outside prospecting for gold, or ministering to clients about their legal rights. She had her hands full with the emotional turmoil of a household on the verge of breaking apart following her daughter's wedding last week. Should she now try to hold the family together?

She was a politician's wife, born in a cabin on Lake Erie in the dead of winter of 1798. The daughter of the United Empire Loyalist, Colonel Samuel Ryerse, she received a land grant of 22 acres in Port Ryerse and set up a homestead there with her new husband John Harris. They were hard years at first since he was a surveyor absent from home for months at a time and knew nothing about farming while at home. After 15 years, Amelia, John and their six children moved to the Eldon House in London where John, now the Treasurer of the District of Upper Canada, had his offices. They lived handsomely on this salary until John's untimely death in 1850. What followed were some of the hardest years in her life. She had lost her first two children, William and Amelia Helen in Port Ryerse, and after the death of her husband of 35 years, her daughter Charlott and her two young grandchildren died in a shipwreck. She had also lost her husband's income. As the matriarch of the complicated family, she scrambled to find ways to keep meals on the table and coal delivered to keep the house livable.

August, 1859 was a difficult time of readjustment for aging Amelia and her family of now-adult children. To solve their mounting financial problems, Amelia had begun the unpleasant task of selling parts of their land, but it would at least be kept within the family since her son's John, Edward and George were the purchasers. Meanwhile, her youngest daughter Teresa had just wed John Scott; a man that Amelia did not think was right for her daughter. Her eldest son John had managed to make a complete mess of his accounting business, which he shared with Edward and George. His reputation was at its lowest point among prospective clients in town. In fact, his clients were abandoning him. He had become irresponsible in handling their accounts, and unable to follow through on even the simplest jobs offered him. Amelia laments, *"There is nothing but care, vexations and anxieties in the world. Oh that I could realize the happiness of a firm well-grounded hope of a happy future"*.

With Miss Loring and John now living in the family home, and frequent visits by the disagreeable mother-in-law Mrs. Loring, the entire spirit of the family and home had taken a turn for the worst. Amelia could plainly see that the confluence of short tempers and ill humor had pervaded the household to the breaking point. Their strength as a family had, most assuredly, never been in their unity, despite the many childhood illnesses and deaths that had plagued this family of 12 children. Under different circumstances, such adversity should have bound them more tightly together. In recent years, however, it had only pushed them farther apart. The center could no longer hold the wheel of the family intact.

Amelia's long talk with Edward her next-oldest son a week earlier had broached the subject of why the family had such a hard time getting along. She was an idealistic woman, especially when it came to her family and how she thought they should carry on with each other. She was steadfast in her opinion that the bickering and fault-finding had to stop. 'If you couldn't find something good to say about someone, then be silent' she would say. 'If we cannot live lovingly together, then we simply have to break up the house and live apart from each other.'

By Friday, August 26th, Amelia came to her decision as the head of the household. It was high time to make the painful decision to ask John to leave the house and find his own home elsewhere. It would be the fall of the first domino that would set in motion events that she viewed with misgiving. Yet, she thought the decision would be best for John's young wife. How was this young wife supposed to establish leadership of her own home if she remained in Amelia's shadows? She saw the inevitable moving out as the beginning of a road she was not so interested in traveling. At the age of 61, her biggest fear was that there would be no one to take care of her when she got old and infirm. Her sons John, Edward and George, together with Theresa and John were living with her now. If she lived here with John and Theresa, then Edward and George would leave and the home would be broken up. In the next few days, decisions had to be made, and there was a nagging feeling that the family was running out of time. The tempest-like lives of the Harrise's this week were far removed from the post-vacation reverie of Simon Newcomb in Massachusetts a thousand miles away.

Young astronomer Newcomb had just returned from a pleasant early-August vacation at Pigeon Cove on the rocky coast of Maine. For a scientist, the concept of a vacation can be a watershed event because there is such a vast distance between the mental state of being 'on vacation' and the intense tedium of being 'at work'. Vacations were a powerful means of recharging creative batteries, and indulging in the simple pleasures of 'play'.

Simon, at age 24 was a lean, muscular and athletic young man who enjoyed afternoon games of baseball, and had a penchant for taking long walks. He had thoroughly enjoyed the long afternoons of strenuous rowing from the Cove to Rockport and elsewhere with his acquaintants, Miss Whiting and Miss Page. In fact, he had spent quite some time rowing with Miss Thining, Mrs. Bullard, Mrs. Batcheliler, Mrs. Jackson and Miss Page too. Evenings were often spent at his favorite game of chess, or engaged in spirited conversations. He was a master conversationalist on subjects that ranged far beyond the technical minutia of a scientific life. Yet, with a career now well-established as 'his first great step in life' he turned increasingly to matters of the heart.

His relationship to Miss Whiting was more than just a passing summer fancy. She lived in Watertown and he had been in her company on several occasions before the vacation. His growing fondness for her had led to a private conversation with her at Pigeon Cove, which he considered the 'second great step' of his life. But for reasons he could not comfortably describe even in his private diary, she had taken this same conversation and its many revelations with remarkable coolness. His very next diary entry on Monday, begins with a series of cryptic inscriptions in a secret code, which was never used again in all the diary inscriptions to come. Whatever he had confided in Miss Whiting in Pigeon Cove, now long since buried in time, it would make a mark on Simon's actions in the weeks and months to follow.

Returning to Boston and after having 'quite a talk' with Harriet his long-time acquaintance, she disappeared from his regular social engagements for some time afterwards, and never took on the same weekly tempo of encounters that it had enjoyed in the months before the vacation. Sometimes she would just appear unexpectedly in his apartment. At other times he would make the journey to her apartment in Boston and not find her there at all. Yet Miss Whiting also vanishes from his diaries even more completely. Her 'cool' reception to his comments while on vacation had taken their relationship in another unknown direction.

The vacationing in Maine had left Simon relaxed and ready to resume his work back in Cambridge at the American Nautical Almanac Office. To most people, his work would not have sounded as glamorous as gold prospecting, lawyering, or mending family bridges. For Simon, work consisted of demanding calculations for new astronomical tables to be used for navigation at sea. For this he was paid $600 a year – a small sum of compensation compared to even a junior telegraph operator. It may have seemed to some like a boring and unprofitable occupation, but it had a lasting value well beyond anyone's common understanding of the hours of tedious work – a fact for which Simon was not completely unaware.

In the years before 1852, American sailors and scientists had been forced to rely on foreign almanacs for predicting the positions of the sun, moon and planets. Through an Act of Congress, the Nautical Almanac Office in Cambridge was set up in 1849 with Prof. Benjamin Pierce the Harvard mathematician as the chief 'computer'. Three years later, the first American Ephemeris and Nautical Almanac for the year 1855 went to press, and was hugely successful. Many people, both scientists and ship's captains alike, came to think of the book as an important demonstration of the developing scientific prowess of the United States. How, then, could this be measured on the same scale as a few grams of gold, or another legal case won at the bench? Like a difficult legal case that stretches for interminable month after month, the calculations for the Ephemeris were so laborious that

forecasters and human computers such as Simon, needed a three-year head start before the results could be printed and published.

Boston Tremont Street ca 1860 (Courtesy Halliday Historic Photograph Company, New York Public Library Digital Gallery)

Returning to the office from Maine, Simon Newcomb re-discovered the doldrums that almost always follow a good vacation. His vacation had been far too successful in reducing his anxieties and pace of work. He tried to move the short mental distance back into the office routines, but his heart wasn't quite up to the effort. He dutifully wrote in his diary on Thursday that his motivation and energy to resume his work had utterly escaped him, leaving him "dull and low-spirited". Today, Friday, his mood seemed much too far from the sharpness he would need to even begin the planning for any mathematical undertakings. In many ways, to produce the long pages of difficult calculations by hand was equal to the taxing effort of running a Boston Marathon. Looking out the window across his cluttered office, the warm afternoon sun beckoned. So he went out for another thoughtful stroll among the tree-lined Cambridge streets and byways. He was caught, like a leaf in the wind, between his complex interests in the women around him, and his research. There was much to mull over between the welcomed emotions of the heart, and implacable logic of the cosmos. Autumn was fast approaching, and these were after all the sunny springtime days of his life. His thoughts turned to where his travels might take him in the years to come; Years that, unknown to him, would ultimately bring him more fame and recognition than anything he could now imagine. But how wonderful it would be to take the rest of this journey with a companion! That wish, however, would require three more turbulent years before Mary Caroline Hassler stepped into his life. While Newcomb strolled the Cambridge Common in search of late-summer's inspiration, another astronomer further on in his career was busy in England with his own ambitious and single-minded labors.

Carrington's Observatory (Courtesy Royal Astronomical Society)

Every clear day for the last six years, the wealthy and middle-aged, Richard Carrington (1826-1875) had patiently opened his private observatory in Redhill, England to let the sunlight stream through his instruments. His father was the owner of the Royal Brewery in Brentford, and by all accounts the family, including his younger siblings David and Esther Fanny, were collectively quite wealthy. After a three-year stint at the Durham Observatory, Richard decided to use some of his family's wealth to build his own manor house and attached observatory at Redhill in 1852. A transaction of family wealth of this magnitude is hard to imagine without a parental death in the family serving as the catalyst. But though his father had preferred Richard study for the ministry, he supported his change of career in college to matters of science and astronomy.

The main house built on Furze Hill on the old Gatton Estate, was a two-story brick colonial with ground floor bay windows facing a fenced-in front yard. The low shrubbery demanded only the lightest of maintenance from a man who had little time for such things. Three fireplaces provided abundant warmth to the occupants. A large lounge room occupied the east-side of the ground floor with a magnificent fireplace and sitting area. Across the central hallway was the west-side Dining Room, Butler's Pantry, Kitchen and Scullery. The second floor held the master bedroom with its fireplace, and guest rooms. An attached single-floor building shared the east wall with the Lounge. It featured a large library and computation area, a central transit room with its sliding roof, and the observatory. The dome of the observatory sat atop the building on its far-east end, and was accessed by a spiral staircase in a hallway leading to the transit room. His assistant, George Simmonds, had two apartments to either side of this hallway. His accommodations sounded luxurious, though they were, in fact, rather small 8x14-foot rooms. Hardly enough space for a bed and chair, but he had access to much of the entire house for reading and work, and his meals were provided by servants.

The telescope was an expensive masterpiece of engineering for the time, built by the master optician Mr. Simms who as Carrington noted '…has not been unmindful of his well-earned reputation' The massive piers of the [transit] telescope rested on a thick slab of York flag-stone, laid on a considerable mass of concrete. The north and south walls were of stone, 14-inches thick, and without windows. The observatory was bounded on both sides by occupied rooms with fireplaces on opposite sides. 'Without windows, the observatory favors the generation of spiders'. One can wonder that the life-threatening, ill-health that dogged Carrington in his later years sprang from the dimly-lit, mold-infested observatory confines he had fashioned for himself at Red Hill.

This 6-inch Alvin Clark refractor was built in 1860 for astronomer S.W Burnham in the United States and now resides at the Washburn Observatory of the University of Wisconsin. (Credit: S.W. Burnham, 'A General Catalog of 1290 Double Stars')

While surveying the springtime Redhill landscape for his new observatory and house, he had taken a walk on Regent Street to acquaint himself with the shops and merchants. It was there that he met and fell in love with Rose Rodway, and even before the construction of The Dome had been completed a year later in 1853, they were married. She was largely unschooled, and came from a very humble and impoverished background. She also brought to the marriage a dark secret; a tempest that was destined to burst upon their domestic scene in a few short years.

Richard often found his solitary studies of the sun interrupted by his responsibilities in running the Royal Brewery. Some of the bookkeeping work could be carried out from the Redhill Estate, known as The Dome to local residents, but all too frequently he had to catch the train to Kew and walk the short mile to the Brewery to administer to tasks in person. The Royal Brewery at 23 High Street lay on the southern side of Brentford not far from the bridge over the Thames River that led to the spectacular Kew Observatory. It was one of the largest breweries in town and produced over a million gallons of ale serving a hundred pubs in its heyday. In the middle of vast barley and corn farms, there was no lack of raw materials for making ale, nor a lack of clients to consume it.

Still, despite these frequent trips, the life of an astronomer was solitary and quiet. His gentle and devoted wife had abundant free time to socialize with neighbors, shop in the nearby town, and attend church functions. Other than the tock-tock-tock of the clock on the wall, and the random clatter of the servants, there was no other sound within the dwelling. For an hour or two around mid-day, he made his solar observations and recorded them carefully. At first, it must have been exciting to view the sun each day to see if a new spot had materialized. Like Christmas, what new blemish would rear its face under the solar tree? As time wore on, however, the novelty abated into a grim sort of drudgery. After a thousand days of meticulous study, one spot tended to look much like the rest. Yet Richard had to stay sharp and not let the tedium dull his faculties for registering even the slightest and most insignificant detail. He was a scientist, and his craft demanded no less.

On Thursday August 26, sunlight completed its journey of 93 million miles, passed invisibly through the aperture of Carrington's elegant telescope, and exited a few meters later onto a large white screen. Richard carefully watched the shadow of the vertical wires that neatly bisected the sun's eleven-inch disk on the screen. In the windowless and often dank observatory room, his expensive chronometer ticked out its measured cadence on the cold stone wall in the background. As a sunspot touched the wire and passed across, he carefully noted the event and its time in his log book. The noisy din of traffic from Redhill's encroaching business district sometimes made the chronometer's tock-tock-tock hard to hear. But despite the din, he had no choice in the matter. His observations demanded that his observatory be fully opened to the sun, and whatever noises might sneak in.

The sun spots, themselves, did not move, but were carried by the imperceptible spinning of the sun across its visible disk – a trajectory that took two weeks to complete from the eastern limb to the western limb. The sun's face had become very complicated since mid-July with a pageant of very large spots appearing and disappearing as the sun rotated in its customary four week cycle. The persistent spot now just over one month old since he first saw it, was complex and peculiar in shape. It had the appearance of a dozen smaller spots clumped together, as though the original single spot had at long last begun to come apart due to forces hard to imagine. He wasn't the only other person keeping watch on this particular spot. That honor also fell to the wealthy Gateshead amateur astronomer.

Robert Sterling Newall (1812 – 1889) had made his substantial fortunes by inventing the wire-rope in 1840, and fabricating twisted submarine telegraph cables designed around the same construction principle. By 1870, Newall had constructed and installed at his estate the largest refracting telescope in the world, a massive 26-inch Cooke instrument, which was ultimately given to the Athens Observatory in 1956. But for now, his modest 6-inch instrument was more than adequate to placate his interest in sunspots. In his frequent patrolling of the sun, Newall had also noted the large spot now appearing on the sun, which he mentioned in a letter to the Editor of the London Daily News.

London Daily News, p. 2 *Solar Spots and the Weather.* "Sir. Mr. E. J. Lowe has called attention to a solar spot, which he saw on the evening of the 21st. This spot made its appearance on the edge of the sun on the 15th, and as the sun revolves in about 25 days, it may yet be visible until the 27th. It has not changed much in appearance since I first obtained a full view of it, and may therefore probably be visible during another revolution of the sun. The enormous spot which is just making its appearance is about 60,000 miles in diameter, while the one referred to by Mr. Lowe is not one-fourth that size. From about 500 recorded observations which I made from April 1948 to April 1952, I am convinced that these spots or holes exercise an immediate influence on the temperature of our atmosphere.. I hope that this notice will elicit information from other observers of solar spots, which are deserving of far more attention than appears to be given to them."

There was nothing else particularly unusual about the spot apart from its size – at least nothing that caught the attention of Carrington's trained eye, or even Newall's. After a thousand days of studying the sun, Carrington had never seen anything more striking than the slow changes of sunspot forms: filamentary shapes slowly changing by the hour; new spots emerging between older faded ones over the course of days. Nothing ever happened, it seemed, in less time than it took to eat a full lunch prepared by the servants, or walk the sort distance to the center of town and back. In the days to come, however, all this would change in one historic moment lasting only a few minutes.

The prelude to this event wasn't to be found in the view from a telescope. Undetected to the eye, or to any magnification by a telescope, the sun had ejected a burst of energy into space that was at this moment on its way towards Earth. Its arrival did not come with a flash of light. Instead, it came as a hidden magnetic message flowing through Carrington's body as he gazed at the unchanging sunspot figures.

Meanwhile, the scientists at the Paris Observatory were used to the unexpected. The magnetic instruments they kept watch on, were always a source of odd events. Most of the time the needles barely moved, preferring to point in a fixed direction where even a simple compass would register North. But for no apparent reason, they would sometimes begin to swing from side to side. The motion was barely perceptible to the eye, but super-accurate micrometers could easily see this agitation. The scientists dutifully entered the readings into their log book.

On this particular day, August 26th, following a long spell of comparative calm, the instruments began to act up between 9:30 AM and Noon. The compass-like needles shifted their positions by nearly a half-degree – an astonishing amount to a scientist; a mere quibble to a navigator. The event, a 'magnetic storm', came and went within a few hours. The cause as with so many other storms, may have been credited to mysterious cosmic particles or to atmospheric electrical currents as was the custom of the time. No one really understood why Earth's magnetism changed in such a sudden way, or what unseen forces could be the cause of it.

Navigators were concerned about these changes because accurate compass bearings were sometimes the difference between arriving at port safely, or being stranded or shipwrecked on a shoal on a foggy day. For a century, scientists had wondered why these magnetic changes happened. Alexander von Humboldt had even given them the moniker 'magnetic storm', as though the act of naming the phenomenon could exert some intellectual control over it. The answer to the puzzle of magnetic storms seemed just beyond reach. So the measurements continued day after day, year after year, at a dozen magnetic observatories around the world. These instruments were the cutting-edge scientific apparatus of the time. Assistants tended them while scientists with their gray beards and stern demeanors gazed at the accumulating data and its quirky changes with mostly blank comprehension.

What the Paris observers couldn't know at the time was that a powerful, but invisible, flare on the sun had let loose a burst of energy. This energy had washed across Earth at the speed of light and upset the balance of charges in Earth's upper atmosphere. Electrical currents had briefly flowed, creating their own magnetism. It was these magnetic changes that registered on the sensitive equipment. A century hence, scientists would knowingly call them Sudden Ionospheric Disturbances, but for now they went, largely, unrecognized.

Most solar flares come and go without further ado. The eruption of this flare, however, caused an enormous puff of solar gases to be launched from the surface by intense magnetic energies now liberated and able to do their work freely. Under a cloak of invisibility, this cloud carried billions of tons of gas at a speed of nearly two million miles per hour. In a few days, a miniscule piece of this cloud was destined to brush by Earth and set in motion a spectacular series of events that would enter the history books for a century.

Sunday, August 28 – El Nino

The cloud of solar gases arrived at Earth on a Sunday afternoon. Its waves of plasma and energy broke violently upon Earth's invisible magnetic shield, compressing its sunward side by thousands of miles. The two polarities for the magnetic field in the cloud, north and south, were like the opposite sides of a coin. Tossed one way, the cloud had a north field with only a minor capacity to affect Earth's own magnetism. Tossed the other, the cloud had a south field, making it a powerful agent for change in Earth's magnetic environment. This time, titanic forces on the sun had tossed a southward 'heads'. An invisible doorway opened up on the sunward side of Earth as the opposing magnetic fields of Earth and plasma intermingled. The solar lines of force connected with their terrestrial counterparts, creating a billion-cubic-mile region of space where magnetic energies and exchanges of particles flowed. The conflagration was held at bay not far from the top of the atmosphere, though the violence of this battle was telegraphed by magnetic lines of force to the ground below, and even into the core of Earth itself. The first hint of the tempest's arrival was registered by sensitive instruments at a remote observatory in Barnaul, Russia.

Barnaul is one of the oldest cities in Siberia overlooking the Ob River, and the home of a major silver mining center. Thirteen noxious factories and furnaces generate nearly eight tons of silver each year, but the city also lives a double life. For the last hundred years, it has been a bustling center for advanced scientific research and cultural life in all of Siberia. The best graduates of the Senior Mining School could even continue their science and technology education at prestigious universities in far-away St. Petersburg or even Moscow itself. One of the most prized, and mysterious, programs at the School was the measurement and monitoring of Earth's magnetic field. Even by the mid-1800's, magnetism was still something of a mystery to most people, though scientists had made breakthroughs in understanding it.

Among the most ardent explorers of magnetism and a world-famous explorer of uncharted continents as well, was Baron Alexander von Humboldt. By 1805, he had also noted these magnetic disturbances and called them magnetic storms' since they caused the same gyrations of his compass needles as local lightning storms would do. Just as Anders Celsius and Olof Hiorter had accomplished nearly 100 years earlier, during a 13 month period, Humboldt and his assistant also made thousands of half-hourly readings of a compass needle. A human 'reader' would peer into a microscope at a needle on a graduated scale, a setup little more than an ordinary compass. At half-hourly intervals, day and night, the position of the needle would be noted. By the 1850's networks of observatories amassed millions of these observations.

Friedrich Georg Weitsch painted Alexander von Humboldt's portrait in 1806 after the scientist's return to Europe from South America. In the background, Weitsch shows Venezuela's Orinoco River, which Humboldt had explored. (Credit: Nationalgalerie, Staatliche Museen)

Using his considerable influence and popularity, and following a two-decade hiatus caused by European wars, von Humboldt acquired the resources needed to set up a number of magnetic 'observatories' in Paris, Freiburg, and later across Russia in the 1830's. Barnaul was one of these lucky centers of magnetic study. After years of keeping track of every magnetic jump and wiggle, they had become pretty good at what they did.

Most of these instruments were pretty simple in design. A compass-like needle was suspended in a sealed chamber, and outside the enclosure, a microscope resembling a surveyor's theodolite was set up to measure the position of the needle. A typical setup could sense a change of less than a hundredth of a degree in the alignment of the needle. Most of the time it pointed steadily in the direction of magnetic North, yet the measurements continued at specified times of the day.

After remaining dormant for days, at precisely 10:00 AM (13:00 UT) the instruments began a slow but relentless march to the end of their measuring scales. Unknown to the scientists at Barnaul, instruments at St. Petersburg showed a similar behavior though starting about an hour later. The world-famous Russian meteorologist

Professor Adolph Theodor Kupffer (1799-1865), who was the Director of the Central Physical Observatory had not seen such peculiar magnetic readings in his memory, '*...remarkable not only for their extent but for their long continuation'*. While the Russian instruments continued their gyrations, instruments in Paris and Helsinki were also beginning to show signs of magnetic disturbances at 5:00 PM (16:00 UT) and at 23:00 UT. Something was unsettling the Earth's magnetic field across thousands of miles. This could only mean that a full-scale magnetic storm was in progress. Yet, standing outside in the afternoon sunlight, one would be hard pressed to see anything but an often cloudy, and unenlightening sky. The disturbances were noted in the customary log books, and the scientists went about their other duties and investigations. For other observers under clearer nighttime skies, the spectacle that awaited them in a few hours was far more exciting than a dancing compass needle.

It was foreordained that Sunday would be a moonless night. The celestial stage had been set millions of years ago, as the movements of the moon and planets played out their choreography under gravity's guidance. It is amazing to think that, had the moon begun the dance a mile further on in its path around Earth, by now it would be full, and lighting up the nighttime countryside. Instead, the dark new moon was fated this day to partially eclipse the sun, but only observers in the Southern Hemisphere would get to see the event. As luck would also have it, clear skies prevailed over most major cities in North America and Europe, except that most of Scandinavia would be cloudy and storm-tossed.

Mary Todd Lincoln finished her morning chores at their residence in Quincy Illinois. She attended church as was her custom, and afterwards decided to sit at her desk to write a letter to her son, Robert. He had just left for Boston a few days earlier to take the entrance exam at Harvard; an exam he was destined not to pass despite everyone's best wishes. A short letter to her dear friend Hanna Shearer soon followed as the afternoon sun made its way to the west. Much had happened in the Lincoln family's life since her earlier letter to Hanna in June. There had been elegant parties and memorable events to fill these warm summer days, but right at the moment Mary was in a melancholy mood over the loss of her eldest child to adulthood, 'it almost appears as if light and mirth had departed with him' she wrote. Like many mothers before her, and others to follow, the empty nest experience was painful and at times hard to bear. As she finished telling Hanna about her son's departure and her emotional turmoil, she could not help but include some tidbits about a more uplifting story: the exhilarating trip she and Mr. Lincoln had taken in early August. It was an unexpected 1,100 mile excursion to look over the property of the Illinois Railroad system in connection with a case that Mr. Lincoln was prosecuting at the time 'Words cannot express what a merry time we had, the gayest pleasure party I have ever seen'. There was also time in this short letter to tell Hanna that '*Your sister Mrs. Parrish has moved to town*

and I am going this very evening to see her, they are some six or seven blocks from here." As the letter closed, she felt a pang of wistfulness at having lost both the immediate companionship of her son and her closest friend. *'What a world we would have to talk about, were we to meet. I shall never cease to miss you. I hope you will not lose all interest in us. For I can never cease to love you. Write soon to your attached friend. Mary Lincoln'*

Mary Lincoln mulled over her losses and her deep feelings for distant friends on this Sunday afternoon, and at the same time in far away in London, Ontario Amelia Harris sat at her desk, deciding what to write in her diary about the day's events. Her horizons had drawn closer over the years spent in Quebec, but as for Mary Todd, she was also dealing with children growing up and moving out, and what this would mean to her circumstances in life. At age 61, a widow, and financially strapped, the arrival of this day is something she had not looked forward to.

After Sunday dinner, John announced that he and Miss Loring would be moving out for a couple of years. Mrs. Loring would be joining them in this new home. Amelia felt, at the hearing of this, a sense of great relief knowing that she would not be asked to help care for the infernal woman! *'...when Mrs. Loring becomes infirm it will be her daughter's bounden duty to take care of her and have her with her. I feel that I could not live with her and I too shall have my daughters to care for me as no one else can.'*

It was a great weight that had lifted from Amelia, although many of her fears still plagued her. She had struggled valiantly to keep her family together in the one house, but it was now time for some of them to strike out on their own at long last. For herself, she could not escape the realization that she had fewer days ahead of her than behind, and she would need to stay with one of her daughters so that she could be taken care of in her declining years. This arrangement was something of an obsession to her. As a widow with a family, she knew how difficult it could be to be alone and elderly. Especially in an impoverished town like London that had known much upheaval since a hard frost two years earlier had killed all of their crops and tossed the town into economic turmoil.

A thousand miles away, Simon Newcomb was making his regular Sunday walk from Cambridge into Boston to attend services at Kings Chapel, though on this particular afternoon he also chose not to meet with Harriet. His encounters with her were destined to become far less regular as their relationship seemed to be relentlessly unraveling. There was also a lot on his mind, not the least of which was his research and his future prospects for continuing employment.

The auroral light show that was to come as twilight ended, came as a complete surprise to everyone, with no sensible warning at all. Only a handful of instruments around the world were recording in their dances that a major storm had enveloped Earth and was now in full swing. The explosion on the sun that had launched this storm a few days earlier had been invisible even to the most ardent human viewer. Its detritus had flowed past Mercury and the orbit of Venus like some vast tsunami, gobbling up space at nearly 50 million miles each day. The advancing edge of the cloud passed across Earth this very afternoon, and shook the planet's magnetic essence to the core. The field's tenuous comet-like tail snapped like taffy pulled to its limit, sending rivers of energized gas speeding into the polar regions of Earth. The particles met their end as they collided with abundant oxygen and nitrogen atoms. In a split second, light was born and colored the sky in phantasmagoric colors and shapes. At first, the afternoon sunshine drowned out the riot of colors now rushing southward across the sky. But as twilight fell, the colors and forms leaped out of the star-spangled darkness, and gasps of awe and amazement rang out from the throats of millions of people across the globe.

Each person saw a slightly different aurora take shape in the sky depending on their location. In some ways, it was like staring into a diamond and seeing the single beam of sunlight reflected from its many planes. At any given moment, far-flung observers witnessed a different dance, but at the same local time, they shared much the same experiences throughout the night. The night sky spanning the entire globe from Russia to California, and from Canada to Cuba exploded into a fireworks of color and motion, more brilliant than a thousand stars – or a million. It would be one of the most spectacular events ever witnessed by humanity in anyone's memory.

Mary Lincoln made her way across Springfield to visit Mrs. Parrish, while Simon Newcomb returned from Boston. Both saw a remarkable pageant of color and shapes moving swiftly across the heavens. Over most of England, the skies were clear following a stormy week, and offered yet another view – one no doubt shared by Richard Carrington had he bothered to keep a diary about the event. Although their diaries of Newcomb, Harris and Browning all read like a rushed attempt to merely remind their future selves that they had experienced the event, other people around the world were not as reluctant to take note of the events and to make their experiences known to history.

The accounts were spread across a dozen time zones from Australia to Asia, Europe, and North America. Some chroniclers were content to merely say they had seen spectacular moving lights and arcs flashing across the skies. They noted this fact without much interest in mentioning the specific times or details of the changing colors sweeping across the skies. Others were careful observers of every variation and color change, even noting auroral positions with respect to specific stars, and times recorded to the second. What we can easily understand from these accounts is that each watcher was deeply moved, amazed, and at times mystified by what they were seeing. Words were often a struggle to find in composing a coherent description of this unfamiliar phenomenon. Whether you were a priest or a criminal, a slave or a millionaire, there was nevertheless a common human frame of reference that bound everyone together as they stood and watched. The events were magnificent, awe-inspiring, mysterious, and for some, terrifying. There was very little attempt made to critically explore what was occurring, mainly because there were as yet no rational understanding of what events were overtaking the observers. All that one could do was to watch with a combination of human incredulity and awe – and hope that these were not portends of some dreadful disaster in the making.

Unlike the countless thousands of aurora that have graced the skies since the dawn of civilization, this exposition was unique because of the many details we know about it; largely the legacy of Prof. Elias Loomis. His passion for aurora, and for their unique magnetic personalities, had begun 25 years earlier soon after graduating from Yale. Like other scientists before him, he was fascinated by the magnetic variations in compass needles. For 14 months he spent his hours between 6 AM and 10 PM making measurements of these mysterious changes in Earth's magnetism. This was one of those rituals of which nearly all serious investigators of aurora had claimed to have participated, much as building a telescope was often a similar rite of passage for would-be astronomers.

Elias Loomis, born in Willington Connecticut in 1811, has been all but lost to the modern literature of the history of science in the 1800's. His obituary in a then-young journal called Scientific American exalted his standing in the community of meteorologists, and declared that '...among the pioneers in the study of meteorology...the name of [Elias Loomis] must ever be classed'. He was very much the Polymath, as comfortable in mathematics and astronomy as in meteorology and genealogy. In his time, he produced over 100 professional articles, and 13 books on mathematics and astronomy, and enjoyed professorships at Western Reserve College and the University of the City of New York, and Yale. In his time, he had worked with Prof. Alexander Twining at West Point to study meteors, and with Prof. Denison Olmstead at Yale to be the first to discover the return of Halley's Comet in 1835. Later in 1848, he was offered the position at Princeton University that had just been vacated by Joseph Henry (1797-1878), who was on his way to assume the duties of Secretary of the Smithsonian Institution in Washington DC.

When he was not conducting research at the forefront of astronomy, mathematics and meteorology, he continued his weekend passion of cataloguing everyone in North America who were descendants of Joseph Loomis. After 40 years of work, his list included no less than 27,000 men and women. Sadly, though he watched the massive tapestry of descendants unfold and weave long threads into the future, his own lineage would end with his two sons Francis (b. 1842) and Henry (b. 1853) who remained childless. When asked by his friends why he had undertaken this exhaustive life-long tabulation he could only reply that he, too, thought it a strange preoccupation and could offer not a single explanation for this consuming interest. Yet, for someone who had spent 14 months patiently recording the position of a compass needle every hour on the hour, 14 hours a day studying auroral magnetism is not such a surprising extension of the same meticulous and relentless personality. Following the death of his wife, Julia , in 1854, Elias was free to isolate himself even more thoroughly in his work from then on.

The 129 accounts of the August-September 1859 Great Aurora assembled by Loomis and published in seven installments in the American Journal of Science and Arts was, at its core, a legacy of events familiar to those who live under cold Arctic conditions, but which were still a surprising experience to observers in warmer, sub-Arctic climates. What many discovered for themselves was that the forms and changes in aurora don't happen randomly. That's why so many people have for centuries reported the same shapes happening at specific times during a display or 'exhibition' as aurora displays were once called.

On August 28th, the auroral activity had already begun under bright sunshine, but of course their brilliance was still not great enough to rival the sun itself. No sooner had the sun set, however, when people began to notice that parts of the sky were awash in a light very different from twilight. The biggest difference being that it was located to the north, not to the west where the sun had just vanished from the sky. What observers would see after this point in time depended on where they were standing. People groped for words to describe the movements and shapes of the lights in the sky. Again, because the

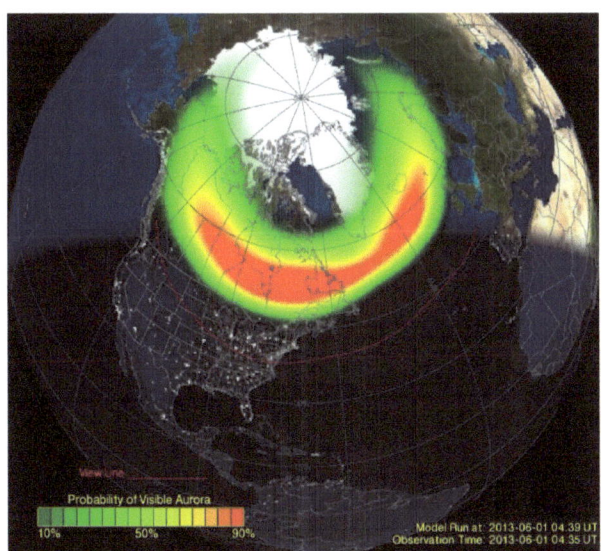

events are not merely random comings and goings of light, observers settled quickly upon a common descriptive language. People spoke of 'beams or rays of light' , 'crowns and coronas', great 'arcs or arches of light' spanning the northern skies. There were also amorphous luminescent clouds, and dazzling red or crimson glows resembling fires. The adjectives magnificent, splendid, brilliant, and remarkable, became overused to the point that they ceased to circumscribe the events at all. At first it is an exciting adventure to read each description, but the excitement quickly blurs into near-boredom as the same, or similar, details are revealed, differing only in the circumstances of time or place. How wonderful it would have been had the diligent, and wordy, observers tried to capture more of their own feelings, or anecdotes about the seemingly extraneous circumstances of the times. Only the newspapers provide much context for who these people were, and what issues of the times were playing across their minds as the watched the iridescent aurora.

The differences in the accounts can be traced to one basic fact about aurora; they are a vast phenomenon in time and space. No single observer ever sees the entirety of an auroral display no matter how long they decide to stay out in the cold on the darkest night. Even a distance of 200 miles can create visions that, like a rainbow, are essentially unique to the observer and to no one else.

Aurora surround the poles of Earth like haloes of light. The first 'quiescent' stage in a display begins with the appearance of a diffuse light and thin streaks or streamers of light. As the display unfolds over the course of a few hours, the halo expands towards the equator and pole, while concentrated bulges of light called 'sub-storms' bloom within the halo and surge westwards and east wards. Within these sub-storms, observers see all manner of curtains, arches and streamers of light, flowing east and west like some heavenly rivers, while at the same time sweeping across the zenith and expanding southward to the equator. Once this middle 'expansion' stage exhausts itself, we are once again returned to a calmer period of quiescent, diffuse light.

Although the August 28-29 aurora followed the usual script and choreography of many aurora that have been seen in the 150 years since then, there were some surprises.

The biggest surprises were the speed with which the ring of light expanded towards the equator, and the breath-taking brilliance of the display itself.

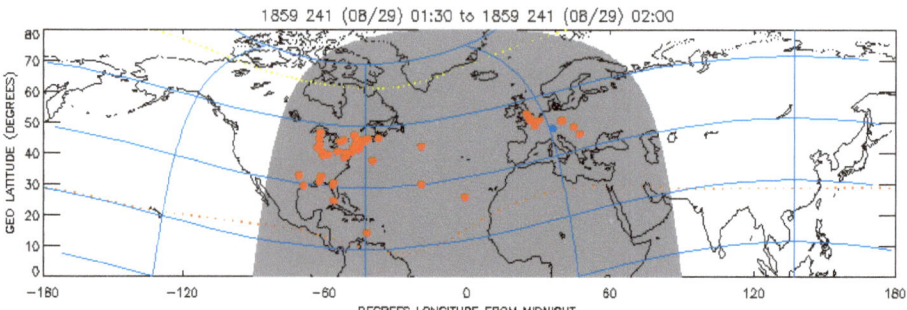

This figure shows all of the aurora observations as of August 29 at 02:00 UT (Green et al., 2006)

From Boston, the aurora began with a rose-colored glow on the northern horizon at 7:30 PM. Within 15 minutes, the aurora had climbed up from the northern horizon and passed beyond the zenith headed south. This would have been a spectacular sight to see, as the aurora spread from horizon to horizon faster than the speediest storm clouds. By 9:30 PM, the expansion of the aurora had crept steadily southward until the entire dome of the sky "was hung with white, gold and rose-tinted streamers". Other observers saw crimson arches, yellow or white arcs, and many other variations on the general design of an auroral curtain, shimmering and coruscating in the mysterious winds of the upper atmosphere. This state lasted for nearly an hour. From Nova Scotia, "between 9:15 to 10:15 PM a diffused light from the entire sky was so bright that it made the landscape easy to see. Cocks crowed and the animal world seemed to think that day was dawning." Yet more incredibly, observers in Key West Florida, Bermuda [Loomis III p.261] and Cuba sighted the auroral curtains and streamers near zenith as well.

Normally, the halo of light around the Arctic regions prefers to stay at latitudes near 65 degrees North. Observers at lower latitudes always see the displays in the northern sky, and distinct shapes are rarely, if ever, reported. Across most of North America to as far south as Florida, only a diffuse, featureless glow of crimson light suffuses the northern sky. What had happened on August 28 in those few minutes after 9:30 PM was that the entire auroral halo had flowed past the latitude of Boston, and had swiftly taken up residence above Florida and Cuba, where it remained for the better part of an hour. Rather than a thin band of light, however, the entire geographic region northward of Florida was covered by auroral forms and spectacular colors.

Auroral corona Québec, Canada Sep. 11, 2005 (Courtesy : Philippe Moussette)

Many reported seeing crowns of light (called the 'auroral corona'), located high up near the zenith, with beams of light shooting out from a common center like a meteor shower. These beams were bright, and exquisitely colorful. They changed in direction, color and luster from second to second like a laser light show painted on the sky of a modern city. What these many observers were seeing was an auroral curtain passing directly overhead. Like the meteors in a meteor shower, or the convergence of railroad tracks towards the distant horizon, the rays from the curtain seem to radiate from a fixed point in the sky.

Usually, auroral displays wax and wane within a few hours and are gone. Yet rather than fade like a fireworks display past its crescendo, the August 28th display seemed to draw energy from the multitudes of viewers. No sooner had it begun to pass its peak brilliance near midnight with its inscrutable energies spent, but an entirely new auroral exposition followed an hour later on August 29th for watchers in England, Boston and Illinois. By some accounts, this second display exceeded the drama of the early evening exposition, and then lasted until daybreak. This changing tempo and variety of shape and color spanning an entire night was far too much for the curious bystander to ignore. What new spectacle would I see if I wait a few more minutes? What will the climax look like? What new forms and colors will I experience? It was, for many, impossible to stop watching the display as though their minds had been overcome by some new force of nature.

Even a busy politician like Orville Browning found the presence of mind to mention in his brief, though profoundly understated, diary entry 'Beautiful Aurora Borealis at night' as he recalled the celestial pyrotechnics of that evening from his location in Quincy, a few hundred miles to the west from where Mary Lincoln was watching. In Canada where Amelia Harris lived and sought inspiration for a troubled family, the spectacle took on a different shape and choreography. Amelia's thoughtful diary note was brief, but captured the essence of the experience nonetheless:

"Last night there was an aurora of unusual beauty which continued all night illuminating everything like a bright moonlight, and changing color from bright red to milk-white. It radiated from the center of the heavens. I got up several times in the night and looked out the windows, and could see the flowers in the garden and move about with ease in my room without a candle, although there was no moon'

Simon Newcomb took a moment from his work, and a nighttime stroll, to note *"A splendid aurora this evening. Rays of red and white flames diverged from a star about 6 or 8 degrees north of the [constellation Delphinus] at 9:55 PM'.* It was hardly the kind of comment you might expect from a professional astronomer and scientist. One would certainly have expected much more from someone who studies the stars. But for Newcomb as for many scientists of the time, aurora were seen as phenomena of the upper atmosphere and meteorology, not cosmic events that directly impinged on their astronomical expertise. As observers, they were no more careful in observing aurora as the average citizen would have been, and in some instances far less attentive.

The sometimes jaded perspective of science aside, how, then, did the Common Man regard these displays? What must have gone through their minds as they watched the sky awash in crimson and glowing clouds of light, which silently moved across the sky? It depended on many things, the least of which was an accurate scientific explanation. In fact, there was none available. Despite the careful studies by a growing cadre of scientists, the average person was still under the grip of many superstitions and folk legends when it came to thinking about aurora. The general reaction at that time to these infrequent spectacles was probably not much better than what was noted by an author of an article in the newspaper 'Flying Post'. On March 8, 1716 following a dramatic aurora on February 23, 1716, *"Some ignorant people, whose ideas are on such occasions stranger than their senses, fancied they saw armies engaged, giants with flaming swords, fiery comets, dragons, and the like dreadful figures"*

The prevailing impression, as it is today, was one of awe and sublime wonderment. It was an hypnotic experience. No sooner had one series of auroral movements completed itself, but another began. No one wanted to turn away from the events of the moment for fear of missing some grand crescendo. So they stood and watched, or periodically checked the sky throughout the sleepless night. The brightness of the display that rendered night time into daylight, made it hard to sleep, and easy to stare out the window for lack of anything better to do.

Perhaps no one better captured into words what this experience evoked than American poet William Ross Warren who found that the aurora were far more angelic than sinister. While the lights continued to play overhead, he wrote;

O ye wonderful shapes
With your streamers of light
Blazing out o'er the earth
From your ramparts of night;
With your strange hazy hues;
With your swift-changing forms,
Light the red-lightning rush
Of fierce tropic storms –
O ye terrible shapes!
Yet through all still appear
Yonder love-speaking eyes
Of the far starry sphere;
So 'mid terror, we still
Can a symbol behold

The aurora exposition was by all recorded accounts, a major public spectacle. Thousands of people in Boston, New York and Baltimore, and millions elsewhere, stood outside their homes, or stopped in their tracks on city sidewalks. With their faces tilted upwards to catch the rare event, they were spell-bound. People looking north from a distance in the direction of major cities, were frequently convinced some horrible fire was raging out of control. In Washington, D.C., the glare in the northern sky was sufficiently vivid to call out some of the fire companies. In Jamaica the aurora was taken for the glare of a fire associated with recent riots. And it wasn't just land-lubbers who were puzzled and concerned by the rare displays.

The seven year old clipper ship Northern Light, arrived at Boston from Africa and was at sea on the night of the aurora borealis. The vessel was struck by lightning twice, after which the red flames of the aurora burst upon the astonished vision of the crew. Most of them were convinced they could smell a sulphurous odor all night.

As stunning as the light show had been, many curious effects also accompanied it at the time, or soon afterwards. It became very clear that aurora had some ineffable personality to them that had not been previously appreciated. While many people were outside marveling at the changing curtains and rays of light, the most remarkable thing about these aurora was the unusual amount of atmospheric electricity that seemed to follow them about.

At this time, printing technology had evolved from cut pieces of wood stamped in ink, to electrotype plates. At the office of the Springfield Republican newspaper in Illinois, the charging of these plates to hold the ink was so seriously affected by the aurora they could not be used to print from so long as the aurora continued their dancing. Even as far north as Montreal, the electrified air rivaled what thunderstorms often yield, causing sensitive instruments to register changes in atmospheric electricity that seemed to keep time with the aurora. Mary Lincoln no doubt received her daily morning newspaper in Springfield a bit late on the morning following the aurora. Meanwhile, in Versailles, France, regular measurements of the amount of ozone in the air also detected a surprising increase at the time of the aurora. There was without a doubt, some undefinable electrical aspect to the aurora. But perhaps the most exciting effect of the aurora beyond its impact on the human psyche, was its effect upon one of the most sophisticated exemplars of human technology at the time: The electric telegraph.

Since its invention in 1837, over 120,000 miles of wire had been set up world-wide, coded messages filled these lines and undersea cables around the clock. Telegraphy had been embraced by politicians, newspaper reporters and the Common Man, to send news and short letters to the four corners of the Earth. With 1500 stations in the United States alone, and five million messages sent annually, it was a technology that had become truly indispensable for running a young nation, or the distant older ones across the Atlantic.

The August 28 storm had been extensive. Throughout the entire northern portion of the United States and Canada, and across much of Europe, Russia and India, telegraph lines were rendered useless for all business purposes. In North America, telegraph stations separated by as few as a dozen miles could not get the message through against all the chatter. Yet, it wasn't just the frustration of getting your message to the other station that challenged the Telegraph Operator during this time, there was also the very real, though as yet unrequited, likelihood of personal injury. In Springfield Massachusetts, J. E. Selden reported that, on the Albany to Springfield telegraph line a flash passed across from the break-key of the telegraph apparatus to the iron frame, like the flame of a jet of gas. It was accompanied by a humming sound, similar to a heavy current passing between two metal points almost in contact. The heat was sufficient to cause the smell of scorched wood and paint to be plainly perceptible. In Pittsburg, Telegraph Manager E. W. Culgan reported that as the key on the telegraph was being used, intense sparks and streams of fire flew between the contacts. Had they been permitted to last more than an instant, they would have fused the platinum points of the key. Meanwhile, the wire in the electromagnet of the telegraph set became so hot that the hand could not be kept on them. The most spectacular, and now legendary, story is told by Frederick Royce: a telegraph operator working in Washington DC. at his station between 8 and 10 PM.

"I did not know that the Aurora had made its appearance until 8 or 81/2 o'clock. I had been working 'combination' to Richmond, and had great difficulty from the changing of the current. It seemed as if there was a storm at 'Richmond'. Concluding that this was the case, I abandoned that wire and tried to work the Northern wire, but met with the same difficulty. For five or ten minutes I would have no trouble, then the current would change and become so weak that it could hardly be felt. It would then gradually change to a 'ground' so strong that I could not lift the magnet. While the Aurora lasted the same phenomena were observable. There was no rattling or cracking of the magnet, as is the case in a thunder storm. I looked at the paper between the arrestors, but found no holes. Philadelphia divided the circuit at the request of New York, and we succeeded in getting off what business we had. The Aurora disappeared a little after 10 o'clock – after which we had no difficulty, and we worked through to New York. During the display I was calling Richmond, and had one hand on the iron plate. Happening to lean towards the sounder, which is against the wall, my forehead grazed a ground-wire which runs down the wall near the sounder. Immediately, I received a very severe electric shock, which stunned me for an instant. An old man who was sitting facing me, and but a few feet distant, said that he saw a spark of fire jump from my forehead to the sounder. The Morse line experienced the same difficulty in working."

The Big One!

Following the spectacle of last night's aurora, most people around the world awoke to a new day that was much as yesterday had been, at least in its outward appearance. No doubt they exchanged stories about what they had experienced the night before over breakfast. At the counters of General Stores in Denver, in business offices in New York, and in one-room school houses in Kansas, the early morning greetings were no doubt followed immediately by 'Did you see the Aurora last night?'. Some of the larger newspapers literally 'stopped the presses' and rushed into print the first stories about the spectacular, world-wide event.

The Baltimore Sun, p. 1, *The Aurora Borealis*, From twilight until ten o'clock last night the whole heavens were lighted by the aurora borealis, more brilliant and beautiful than had been witnessed for years before….The light streaks shot upwards from the horizon and varied in width and length, and changed as long as the phenomenon was visible. It was a grand sight, and was witnessed by thousands of persons, many of whom never saw the like before.

Aurora September 6, 1996 from Fairbanks, Alaska (Courtesy: Jan Curtis)

The Evening Star (Washington DC)   A brilliant display of Northern lights was witnessed from 8 o'clock to half-past 9 last night. The glare in the northern sky, previous to defining itself into the well-known features of the Aurora Borealis was sufficiently vivid to call out some of the fire companies.

The London Daily News. P. 2, *Aurora Borealis*. Early this morning, between twelve and one, a most brilliant display of the above phenomenon was observed extending from the western hemisphere to the north-west, north and north-east, and reaching to the zenith. The appearance in the west was that of a large fire, but in the north and north-east it was of a violet color, and with great brilliancy. This beautiful display lasted for about an hour, and then gradually died away, leaving a serene and unclouded autumnal sky.

San Francisco Daily National p. 2, *Aurora Borealis*. For the first time in several years we had last night a grand exhibition of the 'Northern Lights'. The first appearance was at five minutes past nine o'clock as told by the fire watchman on the roof of the City Hall. The greatest illumination was at about twenty minutes before ten o'clock, when the light was so brilliant that it shone on Telegraph Hill and the upper story and cupola of Wright's building like the reflection of an extensive conflagration. The light, or rather columns of light, were of a deep red hue, and at one time extended from the horizon almost to the zenith. It was a magnificent sight – quite superior to the Chinese fire exhibitions in the theater.

Because of the many time zone differences, the early-morning events reported in Europe between 7:00 AM and 11:00 AM local time were happening at the same time many North American observers were still seeing spectacular aurora under moonless, night-time skies. The Norway and Brussels telegraph disturbances were working their way through the circuitry while Texas morning skies were seeing crimson aurora warring with Polaris and the Big Dipper. And just as dawn's twilight skies were brightening for aurora watchers in Texas, Australian observers were treated to the last few hours of the auroral display just after sunset. But this was not the end of the spectacular display, only a foretaste of an even more amazing exposition of lights to come.

On Monday morning at 9:00 AM (08:00 UT) disturbances in the instruments at the Brussels Magnetic Observatory could still be seen, and for the next hour, they became so severe that they were forced beyond their measuring limits. A powerful pulse of magnetic energy had once again rippled down Earth's magnetic field and delivered a blow to the instruments. Paris Observatory also showed tremendous magnetic activity during the entire morning in near-synchrony with Brussels. Meanwhile, in Paris at the hub of Europe's financial transactions, telegraph connections had still not returned to normal by late morning so that business transactions could be conducted. Telegraph lines from Wurtemburg to Stuttgard were also affected, as were the lines in Norway. At this same time in North America (6:00 AM EST – 11:00 UT), the auroral displays were subsiding and sunrise was greeting the exhausted viewer. Although most places around the world reported no telegraph problems during the afternoon, in Worthenburg, they were still having problems with their lines by 6:00 PM (17:00 UT).

Aurora March 6, 2000 from Fairbanks, Alaska (Courtesy : Jan Curtis)

In England, aurora were spotted at 23:00 UT at about the same time Prof Kingston, Director of the Toronto Magnetic Observatory, reported faint aurora light could be seen at 8:30 PM EST (00:30 UT on August 30) under clear skies, and from Montreal. Archibald Campbell, the Commissioner of the North West Boundary Survey noted from their camp at Simeahmoo in the Washington Territory that a faint diffused light was seen in the north at 9:00 PM (01:00 UT on August 30) and was still visible at midnight.

Meanwhile, Prof Christoph Hansteen in Christianstad, Norway was trapped under poor weather in Scandinavia was mostly cloudy. It was a frustrating experience for him because, so far as his instruments were concerned, the skies had probably been filled with aurora the entire time since Sunday. Only a thick cloud deck prevented him from having a clear view of it. Unlike the instruments at other observatories in Europe that showed magnetically calm conditions, Hansteen's magnetic instruments had not returned to their normal positions and calm state the entire time. There was still auroral activity dancing in the arctic regions, but only scattered views of it could be had from cloud-free places in North America and Scandinavia – and these were in short supply!

The cloud from the sun was continuing to pass Earth and excite a steady procession of aurora, but by now the worst effects of the storm were now hurtling past the orbit of Mars. With no aurora to be found after midnight, the early morning hours of Tuesday August 30 were returned to the ordinary canvass of the myriads of stars, the faint glow of the Milky Way, and the few odd meteors that might happen to flash into view. More newspapers stepped forward to report the events surrounding the Sunday aurora. Readers may actually have grown weary of reading the same graphic descriptions, which were now falling into common patterns of shape, color and tempo. Only anecdotes of collateral impacts seemed to enliven the retelling.

New Orleans Daily Picayune (p.5) *'The City' Change of Weather* '…Towards half past eight o'clock a singular phenomenon took place. The horizon from north to north east became of a deep crimson hue, which expanding slowly, made the sky appear as if lighted by a Bengal fire…At first it was supposed that some great conflagration had taken place on the outskirts of the city, but it was soon recognized that no natural firs could produce this particular hue…Crowds of people gathered at the street corners, admiring and commenting upon the singular spectacle. Many took it to be the sign of some great disaster or important event, siting numerous instances when such warnings have been given. Several old women were nearly frightened to death, thinking it announced the end of the world, and immediately took to saying their prayers. A fat old citizen tremblingly stated that this was the avant courier of a dreadful epidemic like cholera of 1833, whilst a French gentleman pooh-poohed, and gravely assured us that this was the well-known sign of a revolution in Paris, requesting us to make a note of the date.

Washington Daily National Intelligencer (p.1) *'Local Matters'*, The aurora borealis gave on Sunday night one of the most brilliant exhibitions ever observed in this latitude by the oldest inhabitants. The display commenced soon after eight o'clock in the evening, and continued until daylight….Altogether it was an unusually interesting specimen of a phenomenon as yet imperfectly understood. It left us a pleasant and bracing northwest wind and ushered in a beautiful day.

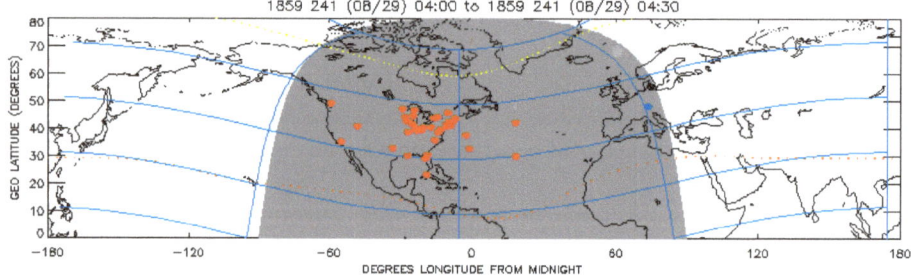

Aurora observations as of August 29 at 4:30 UT (Green et al., 2006)

Chicago Tribune, p.4 *Singular Effect of the Aurora Borealis on the Telegraph Wires.* New York. August 29, The Superintendent of the Canadian Telegraph Company's line telegraphs as follows in relation to the effect of the Aurora Borealis last night: '…so completely were the wires under the influence of the Aurora Borealis, that it was found utterly impossible to communicate between the telegraph stations, and the line had to be closed.' The same difficulty prevailed as far South as Washington.

The London Morning Post, p. 3. *Aurora Borealis.* Yesterday morning a most brilliant display of aurora borealis was visible from soon after twelve o'clock until daylight. Vivid streams of light shot up from the horizon in the north, extending from east and west, which were at times red, and presented the appearance of the reflection from a large fire. The atmosphere was so strongly illuminated that it appeared as if the moon were shining, and rays of light, resembling the rays from the sun as reflected upwards from the back of a cloud, continued to be brilliantly visible during the whole time. The sky was remarkable clear, and the stars shone with as brilliancy that is unusual even in winter.

Daily Morning News – Davenport Iowa, p. 1. *The Northern Light.* Sunday evening our citizens witnessed a beautiful Aurora Borealis. The whole northern havens[sic] were illuminated with brilliant radiations of different colored lights. The streets were lighted up quite bright by them, and the spectacle was a splendid one. The flow of daily events had returned to its usual shape. The only reminders of the amazing auroral exhibition shared by so many people around the world were delivered with the daily newspaper. Even so, most of the breaking news stories were already a day old.

Simon Newcomb offered the briefest of comments in his diary stating that he had sent a letter to his Cousin J.B. Like the plays of Euripides lost to the fire in Alexandria, we know of the letter, but not its contents. Simon had nothing more to say about the aurora within the tapestry of other more pressing events in his life. Orville Browning also focused his diary comments upon the mundane elements of what needed to be done on an otherwise typical day in August. By his diary, Tuesday was cool but pleasant, and he spent it at work in his office.

Such brevity for a diary was most certainly not an issue for Amelia Harris who wrote that '*Mr. Portman offered John a house to live in rent free for one year as a wedding present and we, that is Miss Loring, myself, Mr. Portman and John, went to see the house and I believe John has decided upon accepting Mr. Portman's offer. He thinks by expending 25 Pounds upon it that it will be more comfortable than any house he can get.*' On this eventful day, progress had been made in the Harris family in resolving a difficult matter.

The next day, Tuesday's edition of the New York Times featured one of the lengthiest articles on the aurora to appear in print. Most other papers by this time had relegated the news to a small item of only 50-100 words. This massive article, however, featured many eyewitness descriptions, and even some interesting speculations about the causes of aurora. None of the then-current scientific ideas about aurora and solar influences were mentioned, but several odd-ball ideas did take the stage.

The New York Times, *The Aurora Borealis : The Brilliant Display on Sunday Night,* The present generation have listened with wonder and admiration to the stories their fathers and mothers have told them of auroras and meteors. They have opened their ears and mouths and eyes as they heard of stars falling from the heavens like rain, of the sky at night becoming read as with blood, and in the day time of its being so darkened that stars were visible. Few have had the opportunities of witnessing these sublime displays; but on Sunday night the heavens were arrayed in a drapery more gorgeous than they have been for years…Such was the aurora, as thousands witnessed it from housetops and from pavements. Many imagined they heard rushing sounds as if Aeolus had let loose winds…"

And what of current events? Magnetic instruments all over the world showed no unusual activity, and a near-vanishing of aurora except in the far north where such things were more of a nightly occurrence. The solar cloud had passed Earth, leaving behind a vast region of space that had been 'snow plowed' of any vestige of gas which the sun had previously emitted.

By Wednesday, the story of the Sunday aurora had become ancient history. Only a few newspapers bothered to mention it at all, mainly as a kind of filler for empty space when lucrative advertisements were lacking. Often the newspaper would only reprint articles that had already appeared in other big-city newspapers, rather than waste their own staff reporter's time.

Washington Daily National Intelligencer, p.2, *'The Electrical Light'* [excerpted from the New York Express] The light in the heavens on Sunday night is noted in all directions...The crown above, indeed, seemed like a thrown of silver, purple and crimson being and spread out with curtains or wings of dazzling beauty. Never did the heavens seem to be more the work of the Creator, nor the sublimest work of art sink in comparison so far beneath the wondrous skill and power of the Architect of the Heavens. The tremulous motion of moving light, which the inhabitants of the Shetland Islands call the 'merry dancers' was less apparent than usual, but in place of it came those, full, bright, changing, but more steady streams of light, which gave the intense brilliancy to the whole heavens.

Orville Browning spent another 'fine day' at his office and the evening hours were used to say good bye to one set of guests and accommodate new ones. *'Miss Sue Riddle who has been with us for the last week on a visit, left today. Rev. Dr. Harkey of Springfield, and Mr. Walcher of Hillsborough who were attending Lutherin Synod came to stay with us during the sitting of Synod'*

Meanwhile, Amelia Harris continued to recount in her diary the saga of John and Miss Loring. Amelia seemed anxious to have John and his wife move out, and was becoming concerned that John was being far too picky about which house they would move to, as though they had much financial choice in the matter. At long last, John confided in his mother than they had decided they liked the Griffin's house the best. This was just find with Amelia because she had asked John not to be stubborn and judge the houses the were seeing on his terms, but to take any house that his wife thought would suffice.

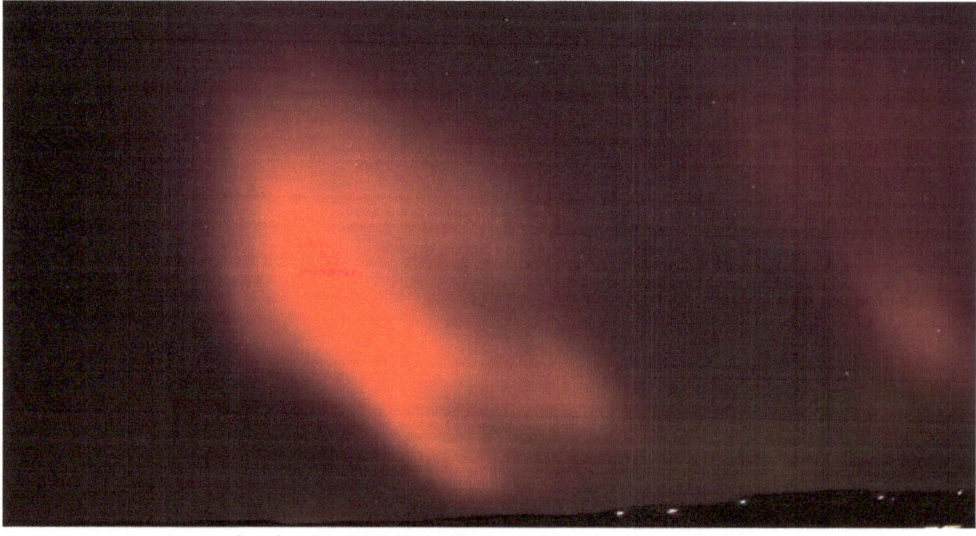

Aurora October 25, 2011 Knox County Ohio. (Credit Joe Golias)

The Deseret News from Salt Lake City published on page 1, a letter to the Editor about the Sunday evening aurora appeared,

> " *On Sunday evening the 28th there was a beautiful and grand display of aurora borealis which lighted up the northern hemisphere majestically and caused many inquiries in the mind of those who witnessed the phenomenon as to the cause which produced it. Much has been said and written on the subject by men who consider themselves learned and wise, but no one unaided by the light of eternal truth, has been or ever will be able to solve the apparent mystery of these remarkable appearances in the heavens, which as many believe, never occurred till after the Ten Tribes of Israel went into the north countries.* "

The Morning Post of London had an article about the aurora today which described a sighting of the aurora of Sunday night from Southampton. The London Daily News also had letter to the editor by Astronomer Newall who reported the Sunday aurora.

An aurora was again sighted in England but in North America, nothing was seen even under Toronto's clear skies. Canada West, however reported aurora just after sunset (7 PM EST) and by midnight the aurora had intensified greatly. It continued to fill the sky throughout the entire night.

By Thursday, the activity was gone from the skies of the US. Observers in Durham and Clifton were the only ones who claimed to have seen aurora that night among the numerous other accounts accumulated during surrounding nights. There were a few newspaper articles who still had something to say.

London Morning Post, p. 5. *Luminosity and Electricity in the Sky*. The heavens were brilliantly illuminated about midnight on Sunday in this neighborhood, says the Manchester Guardian, by a mass of white rays or streaks, completely suffused with a vapor of a pink or dark roseate hue, through which brightly shone the stars, presenting a most beautiful appearance, and being far more deeply colored than the aurora borealis is to be seen in this region. The phenomenon, as seen in Cheale, Cheshire, was sufficiently luminous, notwithstanding some overspreading clouds, to permit the reading of print letters 1-8th of an inch in size…at the zenith a bright and perfect radiation appeared, which extended, slightly interrupted by cloud, a great distance towards every part of the horizon, whilst amongst the rays incessantly played sheet lightning, rendering this the grandest spectacle of that period of the night.

The Evening Star p. 3, *The Aurora Borealis*. Our exchanges very generally speak of the Aurora Borealis which came off on Sunday night last, and unanimously agree in opinion that it was superior in extent and brilliancy to anything of the kind that has been seen in this country for many years. The Superintendent of the Canadian Telegraph Company's lines telegraphs as follows in reference to it: 'I never, in my experience of fifteen years in the working of telegraph lines, witnessed anything like the extraordinary effect of the Aurora Borealis between Quebec and Farther Point last night..'

This, as it would turn out, was merely a brief respite between storms. The next one was about to strike an even more powerful blow. The large sunspot which had launched this storm cloud nearly a week ago, still suffered from magnetic stresses so enormous even last week's paroxysm was not sufficient to relieve them. Another titanic rearrangement of the solar magnetic fields was needed for the stresses to be fully assuaged This event, however, would produce effects even more spectacular than the first.

At 7:18 AM EST (11:18 UT), as the sun rose on another muggy day in Washington DC, the sun erupted in flare of astonishing dimensions. In one minute, enough energy was liberated to vaporize Mount Everest. The magnetic energy was also enough to drive a powerful shock wave almost down to the surface of the sun itself. The light produced wasn't just the ordinary red hue of hydrogen gas heated to incandescence. It was the blinding white color of plasma raised to a temperature of 100 million degrees, making the flare easily visible to the naked eye.

Richard Carrington watched, breathlessly, as a piece of the sun flashed into blinding incandescence. The burst of light startled him as he watched the dark rim of the sunspot. In all the years he had studied the sun, nothing had ever changed at such a pace. The flare brightened by the second and reached its brilliant crescendo as a snaky ribbon of light engulfed a billion square mile patch of the sun. Its violence tore through the rarefied gases at 20 million miles an hour creating a shock wave ten times larger than Earth itself.

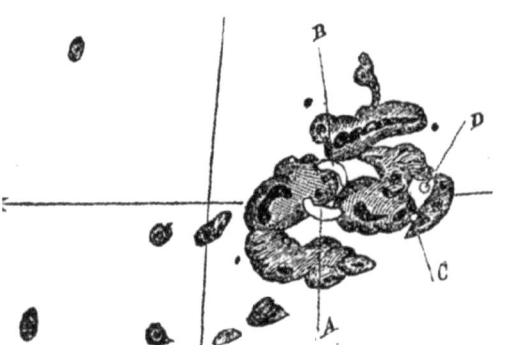

While I was taking my daily observations of sunspots, just before noon I saw something happen which I believe to be exceedingly rare. I was counting from a chronometer and recording the contacts of the sunspots with the cross-wires of my instrument. Suddenly, within the area of the great northern sunspot group, two patches of intensely bright and white light broke out. My first impression was that by some chance a ray of light had penetrated a hole in the screen attached to the object glass of the telescope, because the brilliancy was fully equal to that of direct sun light. I thereupon noted the time by the chronometer, and seeing the outburst to be rapidly on the increase, and being somewhat flurried by the surprise, I hastily ran to call someone to witness the exhibition with me. I returned within 60 seconds, and was mortified to find that it was already changed and much fainter. Very shortly afterwards, the last trace was gone. Although I maintained a strict watch for nearly an hour, I never saw another event take place. The instant of the first outburst was within 15 seconds of 11:18 GMT, and the event disappeared at 11:23 GMT. I thought it would be impossible on first seeing this sudden conflagration not to expect a considerable result in the way of alteration of the details of the group in which it occurred. I was certainly surprised at finding myself unable to recognize any change whatever as having taken place. The impression left upon me is that the phenomenon took place at an elevation considerably above and over the present group in which it was seen projected.

Within five minutes the flash had subsided and was gone. Even the sun itself cannot sustain this level of violence for long. Had Carrington imagined it? Was it a defect in his instrument that had let a piece of filtered sunlight through? Had he delayed his work by an hour, he would have missed the event entirely. Yet the reality of this flare did not solely rest upon Carrington's unique experience of it. Amazingly enough, Richard was not the only human among millions to witness this cataclysm. The wealthy amateur astronomer Richard Hodgson had also spotted the same event at the same moment.

Unlike Carrington, Hodgson had entered astronomy with little formal preparation or advanced training. Born in London in 1804, he was educated at Lewes and entered the professional world, first as a banker on Lombard Street, then as a partner in the publishing firm Hodgson and Graves. He ended his life in these trades in 1841, and thanks to his own personal wealth and passion for science, he first devoted his passions to improving Daguerreotype photography, then took up astronomy. Like Carrington, he built a private observatory in Claybury, Essex, and equipped it with a nearly-identical telescope, which he used for solar observing. While information about Carrington is sparse, for Hodgson, virtually nothing else is known about him other than his obituary and what he had to say in his single published paper on the flare sighting;

> *"While observing a group of solar spots on the 1st of September I was suddenly surprised at the appearance of a very brilliant star of light. The light was much brighter than the sun's surface and most dazzling to the protected eye. It illuminated the upper edges of the adjacent spots and streaks, not unlike the bright edging of the clouds you might see at sunset. The rays extended in all directions and the center might be compared to the dazzling brilliancy of the bright star Vega when seen in a large telescope with low power. The event lasted about five minutes and disappeared very suddenly at about 11:25 AM. From a photograph taken at the Kew Observatory yesterday, the size of the group appears to have been about 60,000 miles."*

Unknown to either astronomer at the time, but in a replay of the events of August 26th, the magnetic needles at the Kew Observatory and the Observatory in Paris, registered almost immediately a major disturbance in Earth's magnetic field at the same moment as the sighting of the solar flare. Again a Sudden Ionospheric Disturbance had been invisibly triggered over 100 miles above their heads, as the energy from this fantastic flare dashed against Earth's outer atmosphere.

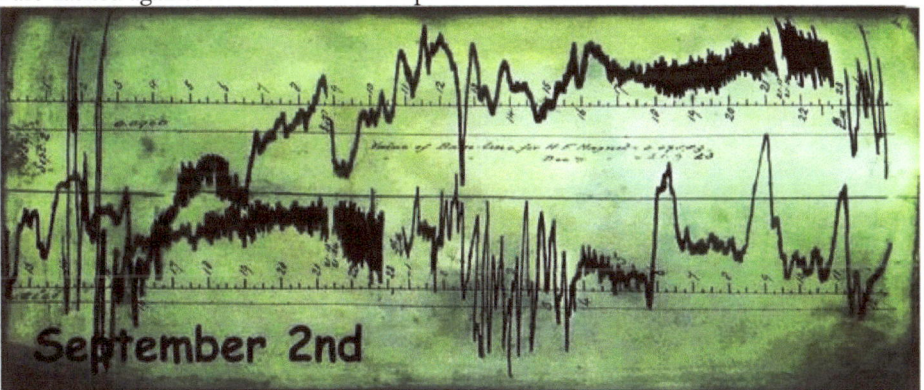

A portion of the Kew magnetogram for September 2, 1859.

Kew Observatory, originally built by King George III in 1769 to study the Transit of Venus, was one of the first observatories to be equipped with a magnetometer thanks to the impetus offered by Von Humboldt. A similar set up was commissioned at the Paris Observatory. Though no one had been watching the sun at that precise moment, Carrington did learn that their magnetometer had caught a curious disturbance at 11:15

AM, about the same instant as the flare. Photographic recording of the data from their instrument had, luckily, begun in 1857 by Balfour Stewart, and not only did they catch this flare event, but also the complete record of the August 28 to September 7 storms. During a one-hour period, the local declination had changed by nearly ½ degree before settling back to its usual bearings.

The Paris Observatory instruments, meanwhile, resumed their disturbed dance at 11:30 AM on September 1 – precisely when the solar flare had been spotted. The magnetic disturbance was an odd, single event, and no further disturbances were recorded. Again, magnetic observatories had recorded a pulse of energy from a solar flare, delivered to the upper atmosphere, causing changes in the ground-level magnetism of earth.

Inspired by the activity spawned by the August 28 storm, Charles V. Walker decided to study the electricity in various telegraph lines at the Ramsgate Station in the County of Kent, England. He carefully noted the direction of the electricity (North or South) and intensity using a galvanometer on each of the lines. His only record for September 1 showed that between 11:20 and 11:26 AM, he had detected a strong current flowing southward in both the 27 ½-mile Ashford and Margate lines. It was the only such disturbance noted for the previous 15 hours, and another like it would not appear for

2000/11/26 08:06

another 20 hours until the morning of Sept 2. Yet the climax of this solar storm was still hours away. As astronomers know today, the most powerful flares almost always pack a second punch.

Coronal mass ejection seen by NASA SOHO satellite on November 26, 2000 (Courtesy NASA/SOHO/ESA)

Moments after Carrington and Hodgson observed the flare, but entirely invisible to them, a piece of the sun broke loose and flowed into space. The superflare had released an enormous amount of magnetic energy. This energy was now finding its way into a larger collection of fields near the sunspot. Within moments, the larger field would become unstuck, and like a balloon released, would float away from the sun, carrying with it 50 billion of tons of super-heated gas. The speed of this event was fantastic, and exceeded seven million miles per hour.

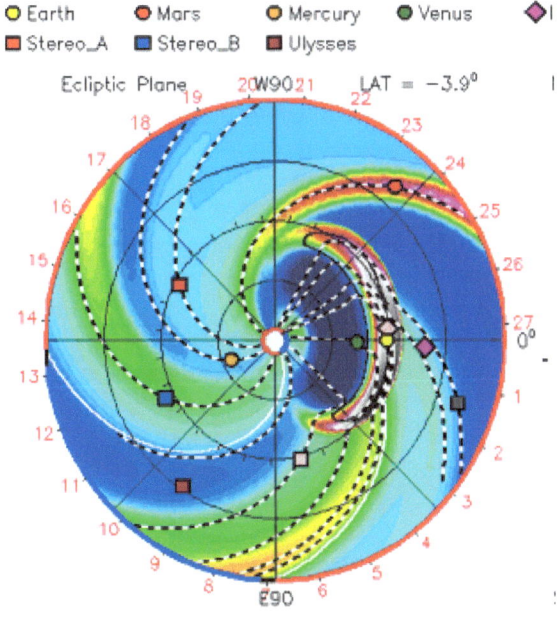

Computer model of a coronal mass ejection reaching the orbit of Earth on January 9, 2014 (Courtesy NASA/CCMC)

The high-speed cloud overtook Mercury in five hours, and Venus in ten. It was an invisible tidal wave of energy, its leading edge frothed and tumbled with magnetic energy and super-hot plasma. It was headed towards Earth, and there was nothing in the solar system that could delay its relentless progress. As it traveled between Mercury and Earth, it moved unimpeded through nearly empty space. The previous cloud had passed this way a week earlier, and had evacuated a vast cavity so that the new cloud experienced virtually no friction in its journey across the solar system.

Carrington originally thought the flare he had seen might have had something to do with the spectacular aurora which followed by his estimate '17 hours and 40 minutes' later. What he couldn't have known at the time was that the storm had begun to show up considerably earlier than this. Even before the first sightings of aurora began to appear, the magnetometers in England and Paris, and even some telegraph lines, had responded to the energies of the Carrington-Hodgson flare, and had been quiet for the remainder of the afternoon. But in Rome, Father Secchi reported a new magnetic disturbance that commenced 3 ½ hours after the flare had come and gone. 'At 4:00 PM (15:00 UT) on September 1, the vertical magnetometer passed beyond its scale, showing a diminution of vertical force.

A second solar flare, though not one visible to Carrington, had been released by the sun spot, and produced a replay of the mornings magnetometer disturbances. Why wasn't this seen in St. Petersburg? The data table published by Loomis (p. 80 Loomis) showed a gap from 12:00 to 19:00 UT, which meant that though data were available, there were no excursions of note according to Prof. Kupffer. Observations at the Greenwich Observatory by Prof. Airy (Loomis, p. 73) show a measurement at 14:00 and 20:00 UT and also miss the exact time of the Rome measurements.

At the magnetic observatory at St. Petersburg, measurements had been made at the unusual pace of once every five minutes rather than the hourly readings. The August 28th aurora began the series, and the remarkable character of the subsequent readings kept the scientists intrigued by the seemingly constant disturbances they were picking up. The micrometer, which kept track of the horizontal intensity of the magnetic field, showed significant deflections during the August magnetic storm, and then subsided to lower levels and minor outbursts afterwards. But on September 1 at 8:00 PM (18:00 UT) they were starting to see the same large disturbances reassert themselves in the data. For the next two hours, the disturbances were stronger than anything they had previously seen since August 28th, and between 9:45 and 10:05 PM (19:45 and 20:05 UT) the micrometer went off-scale. This had not happened at all since they started taking readings.

The start of the severe magnetometer disturbances at 18:00 UT came only 8 ½ hours after the Carrington-Hodgson flare was spotted at 11:18 UT. The cloud of plasma from the sun had finally caught up with the back side of the earlier slow-moving cloud of solar gases that had produced the August 28 aurora. But now, the energies of the two clouds had combined in an enormous wall of matter and magnetic energy, that pummeled Earth's outer reaches like some gargantuan sledgehammer. The magnetometers in St. Petersburg had been the first to pick up this conflagration of matter and magnetic energy that was now swirling around Earth above the atmosphere over 1 ½ hours before sunset. Sunset Greenwich on September 1 occurred at 7:47 PM EST (19:47 UT) The St Petersburg magnetometers recorded their off-scale disturbances beginning at sunset (19:47 UT) and continuing to 20:05 UT. The first sightings of aurora followed within six hours.

At the Colaba Magnetic Observatory near Bombay measurements were routinely made every few hours at preset times day and night, except when a magnetic storm was in progress when the measurements were quickened to 15-minutes. At 7:00 AM Bombay Local Time (is 5.5 hours ahead of UT = 01:30 UT) on September 2, the first indication of a storm condition began to appear, followed by a severe change in the horizontal magnetic force between 10:00 AM and 11:00 AM and a recovery between 11:00 AM and 12:00 AM. What followed from then on was a significant magnetic storm lasting from 12:00 AM September 1 to 4:00 PM September 2.

Shortly before midnight, aurora were spotted over Montego Bay at 10:00 PM EST (02:00 UT on September 2) and an hour later (11:00 PM EST ; 03:00 UT) in Kingston, Jamaica *"It appeared as if there was a colossal fire on Earth which reflected its flames on the heavens. The whole island of Jamaica was illuminated.*

Those citizens who were out and about at this late hour in St. Louis also saw the aurora's maximum brilliance at 11:00PM. This parallels the sightings made at Fort Bridger in the Utah Territory. U.S Army Assistant Surgeon Kirtley Ryland reported a brilliant aurora first seen at 11:00 PM (2:00 AM EST September 2 – 06:00 UT) and reached its maximum brilliance just before midnight. At Cantonment Burgwin, New Mexico, Assistant Surgeon W.W. Anderson, also of the U. S. Army also reported that a guard walking his post at 10:00 PM observed a light reflected from the clouds. It was generally a clouds night, but breaks and thinning out of the clouds probably created just enough of a view of the sky to see fleeting glimpses of the aurora.

Prof. D. Dewey in Rochester New York reported aurora that commenced 'late in the evening and exhibited the usual appearances'. From Newberyport, Massachusetts Henry Perkins saw aurora on the evening of September 1 that were 'quite bright', and which spread rapidly by 12:45 AM on the morning of September 2 to envelop the whole sky.

In Sacramento, California, Dr. Thomas Logan sighted the aurora at 10:00 PM (01:00 AM EST September 2 – 05:00 UT) *'which started as a warm glow in the northwest. Soon the light extended in all directions until the entire firmament was suffused with a ruddy light so bright at times that the hour could be distinguished on the dial of a watch'*

From St. Louis Missouri, citizens saw a beautiful aurora which approached its maximum grandeur after 11:00 PM (10:00 PM EST – 02:00 UT) *"At first there was a hazy appearance, embellished here and there by faint streaks and tremulous touches of light. Then the wavy pencillings grew stronger and broader, and the light spread until it had crept up to the zenith, when half of the world seemed enveloped in a sheet of mellow flame'*

In Galveston, Texas Prof C. G. Forshey saw the aurora commence at 10:30 PM (11:30 PM EST) and continue until dawn the next day.

Whether it was because of a genuine change in the tempo and choreography of the aurora this time, or that observers had become more accustomed as to what to expect, the descriptions offered by eyewitnesses were very different than the ones offered for the Sunday August 28 exposition.

September 2, 1859

The conflagration of matter and magnetism now raging above Earth's atmosphere shook the earth to its core. It wasn't the kind of shaking you might expect from an asteroid impact, or from a stupendous volcanic detonation. The storm slowly enveloped the planet like a gossamer-thin shroud. It pressed firmly on the sunward side of Earth, though the turbulence of its leading edge never quite reached the tenuous realm of the outer atmosphere. It was held at bay by an invisible magnetic field that acted like a compressed spring preventing further intrusion. The sounds of this encounter traveled along the lines of magnetic force. Like a shaken rope, the invisible gyrations and wiggles passed through the atmosphere, earth's crust, and any human being that stood between them.

Pike's Peak miners ca 1859. Note Pikes Peak in background.

Outside Denver City, a small band of miners had pitched camp in an alpine meadow somewhere near Pike's Peak. After a delicious meal of venison and a cup of coffee, they settled down to bed. Only the starry sky illuminated the landscape with its feeble glow. They awoke, groggily, with a well-lit though cloudy sky of an odd and unnatural color – but paid it little mind. Breakfast steamed on the fire drinking their cups of coffee. The fresh brew mingled with the curious aftertaste of dinner's coffee, which was still fresh in their mouths. They packed their puzzled but obedient mules and were set to resume their tedious journey when all of a sudden the sky became dark again. In the confusion of the moment, it suddenly became clear that it wasn't daybreak that had awakened them. Sunrise was still three hours away.

> 'On the night of [Sept 1] we were high up on the Rocky Mountains sleeping in the open air. A little after midnight we were awakened by the auroral light, so bright that one could easily read common print. Some of the party insisted that it was daylight and began the preparation of breakfast. The light continued until morning, varying in intensity in different parts of the heavens, and slowly changing position. We can best describe it as the sky being overcast with very light cirrus clouds, wafted before a gentle breeze, and lighted up by an immense conflagration. It had rained for fifty hours before, only ceasing about twelve hours before the auroral light'

Off the coast of southern Chile, the crew of the steamer Levant had a view of the aurora beginning at 1:00 AM (1:00 AM EST or 21:00 UT). Looking south they saw the Aurora Australis billowing forth in the sky at an altitude of 35 degrees. Soon its fiery red light produced rays of a paler red that shot up to the zenith. Gradually, the whole sky was engulfed in a cloud of crimson light. (Harper's Weekly, 1859 vol. 12/10 p. 787)

Millions of people stopped and stared at the sweeping colors that again filled the twilight sky, re-igniting old fears of Armageddon, and new fears of fiery catastrophes in the making. It had been the second such spectacle in less than a week. What did it all mean?

There were no atmospheric detonations, no shaking of the ground, or gale-force winds from the heavens. Instead, the conflagration that had again turned the midnight sky to daylight, marched onwards for hour upon hour – utterly soundless; utterly alien. The eyes saw fire and luminous clouds, but the ear heard nothing at all. You expected to hear the rushing of winds as the clouds swept past in their flowing crimson colors. You expected to hear the crackling of flames as the clouds changed from moment to moment and rays of light strobed the sky. Yet only the sound of your racing heart and rapid breathing filled your consciousness.

In a replay of August 28, the storm swept across the globe in a vast arc from Asia and Europe, and across the Atlantic to North America. Everyone who had a clear sky was astounded by the spectacle. Most people noted the event, and no doubt spoke of it with incredulity at the breakfast table with bleary, sleep-deprived eyes. Other's watched the riotous colors warring in the night with a critical eye. Again, as they had dome a week ago, enthusiastic observers wrote long and detailed letters to newspapers and journals the next day, describing every turn of color and every change of form. It was as though they were writing a sonnet or a prayer, or counting the minutes to some final climactic event beyond human comprehension. And just as this first storm had passed, another one followed from the ashes of the first a few days later; even more magnificent; even more intrusive.

For the second time in less than a week, viewers in Kinston Jamaica saw the crimson lights appear in the northern sky on the night of September 1st and the morning of the 2nd. It was as if there were a colossal fire on earth that reflected its flames on the heavens, illuminating the entire island. It continued until 5 AM and looked as if Cuba was on fire, and many believe that a portion of this island had been destroyed by a conflagration. Meanwhile, George F. Allen, from Cohe Cuba, noted that *"a Spanish mechanic who worked for me called me out of bed to see the great light in the northern sky. He was much struck with it, and said the people in St. Jago de Cuba would think the end of the world was at hand'*

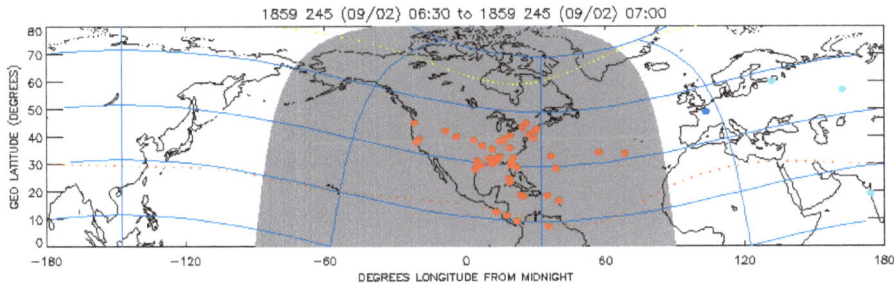

Aurora observations on September 2 at 07:00 UT (Green et al., 2006)

Spectacular red aurora have been seen in many places, but rarely in places that are far from the magnetic poles. Although Japan is at the same latitude as central United States, it is so far from the magnetic pole that sighting aurora from this location is a once in a lifetime event. When seen, they are usually of the crimson or deep red variety. The most intense of these are called 'Sekki' which means 'the red atmosphere'. In Wakayama (then 'Kii') Japan, a Sekki was seen and recorded began at 7:00 PM (08:00 UT) on September 2 and lasted five hours until midnight.

Red Aurora photographed on October 24, 2011 in Independence, Missouri. (Credit Tobias Billings/NASA)

Simon Newcomb, meanwhile, was preoccupied by several mundane errands: *"Went to Boston this afternoon to see about selecting a pitcher for Prof. Winlock. Bought flannel for a shirt, and three pairs of socks made from half wool. I saw another curious aurora this evening"* He had spent more words discussing wool socks than the aurora. Orville Browning mentioned the spectacle in an off-hand manner, preferring instead to note the debate between Warren and Baker on a theological matter ' Another fine exhibition of the Aurora Borealis last night. Pretty day. But so cold I have fire in the office. At night I attended a discussion in the City Hall between Rev. Dr. Warren of the Presbyterian O. S. Church, and Mr. Baker an infidel, of the question *"Has the influence of the Bible been good or evil?'…The debate was conducted with great decorum and is to be resumed tomorrow".* Had he glanced at the Quincy Daily Whig on September 3 he would have spotted the briefest of notes about the aurora on page 2, *"Aurora Borealis – From 12 to 3 o'clock yesterday morning (September 2) , the sky was again lit up with the beautiful Aurora Borealis. Weather-wise sages say that this phenomena presages an early frost'*

The clearest and most detailed account of the aurora was offered by Prof. William Rogers in Lunenberg, Massachusetts.

> *On September 2nd a clear sunset was followed by a peculiar greenish and purplish light extending round the horizon even beyond the north. Over the northeast quarter, the air to the height of 30 degrees had a dark opacity which had the effect of arresting the light coming from beyond. At 7:30 PM, an irregular obscure space began to form along the northern horizon. At*

7:50 PM a faint arch of white light made its appearance, resting on the horizon a little north of the east and west points, and culminating some distance below the pole star. This continued to rise until 8:00 PM when its apex was within a few degrees of the pole. At 9:20 PM a low luminous segment showed itself on the horizon beneath the arch. The latter now resolved itself into an array of bright streamers with equidistant shadowy space between them.

At 9:30 PM the streamers had grown brighter while the low luminous segment, diffusing itself upward, had merged into the outer arch, which now reached nearly to the pole star. At this moment the arch began to send off successive waves of light, rapidly following one another towards and beyond the zenith. In a few seconds this wave movement gave place to more rapid and seemingly broken pulsations flitting upwards in close succession through the northern, eastern and western quarters of the sky, and visible though less distinctly in the south. This wonderful appearance exhibited everywhere a convergency of lines of motion towards a point considerably south of zenith.

When these luminous phenomena were at their height, every spot to which the eye was directed except the southern quarter near the horizon, was traversed by quickly successive flashes of white, greenish and pale roseate light, all seemingly moving upwards.

At 10:30 PM, the pulsating movement again extended over all the northern and part of the southern half of the sky. Innumerable waves of white, yellowish and purplish light chased each other from every quarter towards the magnetic pole, while the crimson flush spread wider and wider from the west.

The various phases of this aurora recurred according to a somewhat uniform order of succession. First the dark segment on the northern horizon took a regular arched form, ad as it rose, became bounded above by a broad luminous curve at the same time developing one or more bright concentric arches within. The streamers now shot forth from all parts of the luminous zone and as these increased the upper arch faded away as if it had expended itself in producing them. And now the lower arch took its place to be obliterated in its turn by a like seeming process of exhaustion. At length, one of the grandeur effusions of light coming on, the whole arch was broken up and the dark segment below was reduced to a shapeless mass. Then there occurred a comparative pause in the phenomena until the dark segment again took form with its one or more luminous bands and a like cycle of development was repeated.

From Preston, England there was a brilliant auroral display continuing from 11 to 12:00 PM and a second appearance though not so brilliant at a little before 2:00 AM on September 3. During the first display the whole of the northern hemisphere was as light as though the sun had set an hour before, and luminous waves roiled up in quick succession as far as the zenith, some of a brilliancy sufficient to cast a perceptible shadow on the ground. To the northwest there was a large patch of light of a deep crimson hue, while the waves of light were white as also were the streamers which occasionally shot across the northern part of the sky. It was the most brilliant aurora that has been witnessed here for many years.

The telegraphs in Paris were almost constantly in motion from 4:00 AM to 8:00 AM on the morning of September 2 although no aurora was noticed that evening. Business was again interrupted, the needles were disturbed, and the bells were rung. At 7:00 AM, bright sparks were noticed on the conductors of the lines from Bordeaux and Toulouse.

The aurora were also seen at Athens by astronomer Julius Schmit:

"Sept 2nd 7:15 AM, beginning of a storm from the west, rain and thunder; at 8:30 AM rain hail and lightning. From Noon to 12:40 PM violent shower from the west, then became clear with sunshine. The evening was clear and in the north there appeared a dark bank of ordinary clouds (not the dark segment of the aurora) above which from 7:30 PM to 8:00 PM was seen a fine aurora of a carmine red color. The cloud bank, which extended 60 degrees in azimuth was elevated somewhat above the horizon so that stars were seen beneath it. The center of the auroral light was not in the north but north-northwest. On the west it was bounded by Cor Caroli and on the east by Alpha Persi. No streamers of fluctuations of light were observed. By 10:00 PM the dark bank disappeared and the auroral light having disappeared previously."

Because red aurora were seen almost to Earth's equator, high-energy electrons were being dumped into the upper atmosphere at a fantastic rate. The only large reservoir for such particles on the night-side of Earth are the ring current and van Allen belts.

C. Westbrook, Baltimore Telegraph Superintendent 'On the morning of September 2 I found the telegraph wires charged to an extent far beyond the strength of our ordinary batteries. Upon disconnecting the batteries I got clear and distinct writing from Cumberland distant 179 miles. When the current was at its maximum strength, the manipulations of the operator at Cumberland worked the armature of the relay magnet here. The intensity of the spark at the instant of breaking the circuit was such as to set on fire the wood work of the switchboard."

The disturbances in the Rome magnetic instruments subsided by 4:00 PM EST (20:00 UT), but in another few hours Norway would start to see aurora just past midnight local time (23:00 UT) During the next four hours, aurora would also be spotted in Bloomington Indiana, Montreal and Canada West from 8:00 PM to 11:00 PM EST.

The next day, September 3, the leading edge of the fast-moving cloud and the back edge of the slow moving cloud were now past Earth, and only the back of the fast-moving cloud had yet to fully pass us by in the next day or so. In its transit, it would still provoke auroral storms and disturbances in its wake.

Prof. Alexander Twining from West Point New York noted that

The repetition which took place Sept 3, although on a vastly diminished scale of grandeur, I observed about one hour - say from 9 to 10:00 PM. It was remarkable for the character of the auroral waves, which passed upward, illuminating successively different definite spaces in their path. The motion of these waves was far more moderate than I have ever before remarked. In this instance I could not estimate it to exceed forty-five degrees of arc in a second of time. The movement was everywhere directed upward; but the determination of arcual or angular motion in this phenomenon is excessively difficult and inexact."

Prof. Schmidt in Athens on September 3, noted that the entire day had clear skies, and at 4:00 PM he went onboard a steamer for Syra. From 7:30 to 8:00 PM near the island of Egina he saw in the north and northwest the carmine red light of an aurora. From

9:30 to 10:30 PM near Cape Sunium a faint trace of the aurora was still seen, but no dark segment, streamers or fluctuations of light". Observers in England, Toronto, Canada West and Rochester all had favorable views of the aurora between 7:00 PM and 11:00 PM EST this evening.

By September 4, in the minds of some people, the spate of spectacular aurora spanning an intense one week interval took on much more than a simple diversion from normal daily affairs. The Columbus, Ohio Statesman newspaper had run a short article about a sixteen year old girl ' of considerable intelligence and prepossessing appearance', who had been taken into custody by the Sheriff of Ottawa County. Her agitated state

necessitated that she be moved to the lunatic asylum. The conclusion drawn from this, and no doubt her utterances, implied that she had become deranged from viewing the aurora borealis a short time ago. She was convinced that all of this spectacular auroral activity meant that the world was soon to come to an end.

The Scream by painter Edvard Munch. Ca 1893

Prof. C. Lyman noted that on Sunday evening *'there were indications of a bright aurora, though a clouded sky prevented it from being particularly observed'*. Meanwhile Christoph Hansteen, Christiana Norway reported that *"10:00 PM, radiating aurora in the north to 30 degrees altitude. Later in the night, vehemently flaming with broad flames"*

In Vienna, Prof. Haidinger reported that at the magnetic observatory, great magnetic disturbances were seen in the morning through the evening of Sept 5th. In St. Petersburg, considerable magnetic activity was also still continuing. James Hubbert in Grafton, Canada West, reported that there were no further auroral sightings by this time from his location, though from the Toronto Magnetic Observatory, faint auroral light was still visible this evening, with occasional streamers. However, with the skies starting to clear at last, in Norway, very bright aurora were seen between 10:00 PM and midnight. This was preceded at 9:15 PM by a large disturbance in the magnetic instruments about as big as what Hansteen had seen from his instruments on September 2. Indeed, some remnants of this latest storm seemed to be present in auroral activity that was visible until September 8 in some locations such as Bloomington Indiana. Eventually, however, the two great storm events of August 28 and September 2 blended back into the normal level of auroral sightings from northern locations by September 5, though people scattered about the world were still on the lookout for the next spectacular displays that, ultimately, never came.

Deus Ex Machina

By many accounts, the aurora of 1859 were the most spectacular ever witnessed. For the next 150 years this solar storm and its consequences would frequently be mentioned in popular descriptions of 'space weather' and in professional textbooks on geomagnetism, space physics and astronomy in the 20th century. But for the most part, this vast collection of 100's of thousands of measurements was a frustrating resource for scientists to sift for clues to the aurora's secrets. Professor Joseph Loverings at Harvard, only a few short years before the 1859 aurora lit the skies, he concluded that *'If we look directly at the crowded columns of numbers which record this great mass of observations, we see nothing but chaos, and the clearest mental conceptions, unless superhuman, are unable to trace the law and harmony behind the apparent confusion'.*

Prof. Denison Olmstead (1791–1859) at Yale University came to a simple conclusion about aurora: their causes had to be extra-terrestrial. Olmstead was a contemporary of Elias Loomis who meticulously assembled the records from the 1859 aurora. Olmstead is credited with giving birth to meteor science after the magnificent, and terrifying, 1833 Leonid Meteor Storm over North America spurred him to study this phenomenon. He subsequently demonstrated through parallax measurements that meteors are not an atmospheric phenomenon at all, but cosmic in origin. He died on May 13, 1859 only a few months before the August 28 aurora blazed forth around the world.

So, by 1859 there were at least a few plausible ideas by scientists that proposed aurora were caused by solar interactions with Earth. The problem was that there were no influential scientists of the times to promote these new ideas by virtue of their considerable reputation and personal influence. To make matters worse, many of the most influential physicists of the time working in astronomy, electricity and meteorology were either disinterested in aurora – messy phenomena laced with mythology and a raft of crazy explanations, or could not accept an extra-terrestrial origin for them at all.

Lord William Thompson Kelvin (1824-1907) was a man of incredible self-confidence, and is responsible for more outlandish utterances and pronouncements than most scientists of his time would probably be comfortable making in public. The tendency began as an undergraduate. His ego provided no option for him than to consider himself First Wrangler at Cambridge long before the results of the qualifying Tripos exam were even known. Legend has it that after taking the exam he asked his servant, "Oh, just run down to the Senate House, will you, and see who is Second Wrangler." The servant returned and informed him, "You, sir!". Another example of his hubris is provided by his 1895 statement "heavier-than-air flying machines are impossible". Kelvin is also known for an address to an assemblage of physicists at the British Association for the Advancement of Science in 1900 in which he stated, "There is nothing new to be discovered in physics now. All that remains is more and more precise measurement." Although he was the 'discoverer' of Absolute Zero and a number of important concepts in thermodynamics, some of his detailed calculations turned out to be wrong in other areas. For instance, he calculated the age of the sun as 100 million years. This meant the sun was much younger than geologist Lyell or biologist Charles Darwin required in order to shape mountains and evolve species of life on Earth.

Arguably by some accounts, Lord Kelvin's 'contribution' to auroral science was to destroy the credibility of the Norwegian physicist Kristian Birkeland (1867-1917) who had created one of the best, and most modern, theories of solar-terrestrial influences of the time. In 1898, Birkeland came up with the idea that beams of particles from solar sunspots were striking Earth and causing the aurora to glow. In Europe, Birkeland was widely regarded as having made a fundamental breakthrough to understanding why it was that some sunspots seemed to trigger aurora. Lord Kelvin believed, without proof, that magnetic energy from the Sun could not possibly produce influences at the distance of the Earth. In Britain, the Royal Society was the prevailing stamp of approval for scientific thinking. Because Lord Kelvin was an influential figure at the Royal Society, his view of the impossibility of solar interactions with Earth was upheld through several generations of the Society's leaders. This had the chilling effect that Birkeland was, largely, dismissed as a crank by one of the most influential groups of physicists in the world. Worse still, his accomplishments disappeared from view in the scientific journals in the English-speaking world.

The odd fact is that both Birkeland and Lord Kelvin were correct. Solar particles did stream from the sun, we would later discover, but they weren't in the form of Birkeland's electron beams. Birkeland Currents, or Field Aligned Currents as they are now called, are actually produced within Earth's magnetic field. Lord Kelvin was also correct, because magnetic induction of electrical currents in Earth's atmosphere by the distant solar magnet, would not work out according to Faraday's electromagnetic induction theory, even though the influential physicist Joseph Henry had favored a similar idea in 1860. The numbers just wouldn't work out in any physically reasonable way.

The correlation between specific sunspot sightings and an aurora or magnetic storm, was not perfect. There was something missing; a compelling smoking gun that could be seen to repeatedly operate time and again to cause aurora to begin their aerial dances. One would have thought that an astronomer seeing a flare, and a half day later the advent of a global aurora, would have convinced even the most ardent detractor that specific solar conditions do matter to auroral events. In fact, the 1859 flare did nothing of the kind. Many careful studies were made of the solar surface and magnetic storms after the dazzling flare of 1859, but no other sudden brightenings of the surface were ever identified as the cause of any of the major 'Great Aurora' events to come. Carrington's flare was a fluke. Even Carrington made the remark 'a single swallow does not make a summer.' The general scientific consensus for decades afterwards was that the Carrington-Hodgson flare had nothing to do with the onset of aurora. It was just a lucky, but irrelevant, solar event.

The Astronomer Royale Ellis acknowledged Carrington's skepticism and noted that the sun has been constantly watched for a dozen years afterwards, and a continuous photographic record of magnetic storms since 1857 had been conducted. No conditions similar to the Carrington-Hodgson event were ever seen again. In fact, much greater magnetic disturbances had come and gone, but never identified with a solar flare event of any significance. No one ever doubted that the 1859 event was not of some scientific value. The problem is that its singular nature made it hard to decide just what that value might be.

Despite the problems with working out detailed theories for aurora and their origins twenty years or more after the 1859 storm, astronomers were generally making increasingly comfortable public comments that aurora and magnetic storms seem to be synchronized with the presence of sunspots, and to the newspapers, even though the exact mechanism for this synchrony was still, by many accounts, a complete mystery. The connection between magnetic disturbances and auroral displays was pretty well established

in the literature of the times as well as the idea that aurora are some type of electrical disturbance in the upper atmosphere. The similarity between auroral lights and the glows from artificial, high voltage discharge tubes were also turning out to be more than just laboratory curiosities being studied by Prof. De la Rue in 1881, and re-discovered over 30 years later by Kristian Birkeland with his 'Terella' machine (see photo).

By bringing magnets near these glowing discharges, it was possible to create patterns in the glowing gas that looked like many of the classical auroral forms that had been recorded for centuries in eye-witness accounts. Spectroscopic studies also showed that the light from aurora was created by excited oxygen and nitrogen atoms giving off their pure light energy at specific wavelengths only. Aurora were not produced by burning incandescent particles of matter, but by some process involving atomic interactions at a very elementary level.

The 1859 storm quickly became the standard against which people would calibrate their personal observations of 'Great Aurora' for the remainder of the 19th century. It would not, however, remain the high-water mark for communications disruptions. Relentlessly, more telegraph lines were installed, followed by expanding networks of undersea cables, telephone lines, then wireless communications channels. With each new aurora and magnetic storm, even minor solar 'squalls' would set in motion a pattern of disruptions that steadily surpassed the level experienced in 1859. This meant that the economic impact of these storms was on a steady rise as well. Despite some of the confusion in the late-1800's over the causes of aurora, it was now clear from the vast network of impacts in the wake of the 1859 Superstorm, that aurora had direct real-world consequences far beyond merely jostling a compass needle. The subject wasn't just something to occupy the attention of academic scientists. Aurora, or the invisible agents that accompanied them, could now reach down into the very guts of our growing and evolving communication technology and cause tremendous problems. The intimate, though often inscrutable, connection between the sun and aurora was a lesson in applied physics that we re-learned many times in the 100 years that followed the 1859 storms. No fewer than 37 major solar events triggered significant and expensive disruptions in our technology between 1859 and 1960. Even in the Space Age we continue to be plagued by the impacts of severe solar storms.

Possible first sketch of a coronal mass ejection during the July 18, 1860 total solar eclipse. From Torreblanca, Spain. Drawing by Temple / Raynard

Astronaut Joe Acaba photographed the aurora australis (southern lights) from the space station from 240 miles up on July 15. The Canadarm 2 is in the foreground. Acaba saw the corresponding southern version of the aurora that many of us in the northern hemisphere saw from the ground. (Credit: NASA)

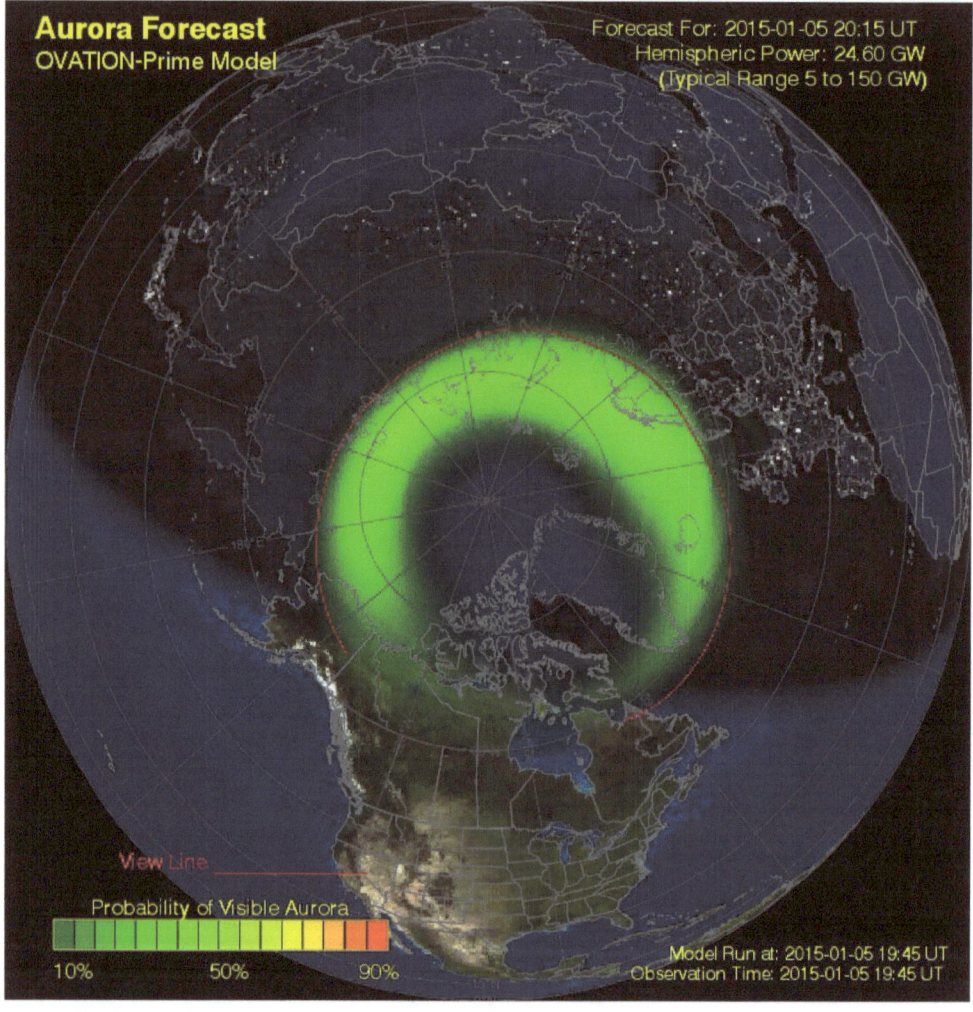

Predicted location of the auroral oval in the Northern Hemisphere for January 5, 2015. (Credit: SpaceWeatherLive.com: http://www.spaceweatherlive.com/en/auroral-activity/auroral-oval)

4 Other Storms of the 1800's

The 1862 Civil War Aurora

December 14, 1862 - This aurora was seen by Civil War soldiers in Fredericksberg Virginia. According to a letter by Milo Grow "There was a brilliant exhibition of Aurora Borealis soon after dark last night. For half an hour it shows very brilliantly reaching to the mid heavens in colors of yellow and red. And in the David Ballenger letters "On the night of the second day of the battle there was a singular appearance in the elements, the most singular that I ever saw in my life. Some said it was an Aurora Borealis, or Northern Light, but if it was it was a little different from any I ever saw before. It rose on the side of the enemy and came up very near parallel with our line of battle, and right over us. It turned as red as blood, but when it commenced rising it looked more like the appearance of the moon rising than anything else I know to compare it to" (1862 December 23). " And by John W. Thompson, Jr. "Louisiana sent those famous cosmopolitan Zouaves called the Louisiana Tigers, and there were Florida troops who, undismayed in fire, stampeded the night after Fredericksburg, when the Aurora Borealis snapped and crackled over that field of the frozen dead hard by the Rappahannock ..." And also in 1905 by Elizabeth Lyle Saxon in A Southern Woman's War Time Reminiscences " It was near this time that the wonderful spectacle of the Aurora Borealis was seen in the Gulf States. The whole sky was a ruddy glow as if from an enormous conflagration, but marked by the darting rays peculiar to the Northern light. It caused much surprise, and aroused the fears even of those far from superstitious. I remember an intelligent old Scotch lady said to me, "Oh, child, it is a terrible omen; such lights never burn, save for kings' and heroes' deaths."

In the Washington Post, January 4, 1885 (p.7) an article 'A night on the fatal field of Fredericksburg – The terrible sufferings that followed a hard day's fighting, among the dead and dying, the graphic story of one who was there' we read 'These are anniversary days. Twenty two years have passed since Fredericksburg. Of what then was, not much is left but memory. Faces and forms of men and things that then were have changed, perhaps to dust. New life has covered some; the rest look but lingering farewells…Splinters of boards torn by shot and shell from the fences we had crossed, served as headstones, each name hurriedly carved under brief match lights, anxiously hidden from the foe. It was a strange scene around that silent and shadowy sepelcure (sic). We gave them a starlight burial it was said; but heaven ordained a more sublime illumination. As we bore them in dark and sad procession their own loved North took up the escort and, lifting all their glorious lights led the triumphal march over the bridge that spans the worlds – an aurora borealis of marvelous majesty. Fiery lances and banners of blood and flame, columns of pearly light, garlands and wreaths of gold, all pointing upward and beaconing on. Who would not pass on as they did, dead for their country's life and lighted so by burial by the meteor splendors of their native sky?"

Aurora painting by Frederic Edwin Church ca 1865. Some speculate he was inspired by the 1859 Aurora.

October 14, 1870 - *Aurora last night: Its remarkable brilliancy* [New York Times, October 15, 1870, p.5]. The most brilliant display of northern lights ever observed in this section took place this evening at o'clock. They were of a deep crimson color and continued about an hour. In Washington DC, the whole firmament was one blaze of crimson light lasting for several hours. The auroral display was seen in New Haven, Connecticut where the red color was mistaken for a conflagration and someone sounded the fire alarm and the Department turned out. The display was witnessed by thousands of people in Boston, Springfield and Worcester, Massachusetts as well as in Norwich, Connecticut. The aurora was also seen in Terre Haute, Indiana around 7 o'clock.

THE RED LIGHT IN THE SKY, OR THE AURORA BOREALIS.

An auroral display [New York Times, October 25, p.1] From the 1868 book called The World at Home, or Pictures and Scenes from Far-Off Lands, illustrated by Mary and Elizabeth Kirby

October 24, 1870 - Cleveland and Cincinnati see a 'splendid' aurora. Widely observed throughout the country, the aurora lasted two days. Various scientific causes are mentioned, with quotes by Prof. Olmstead. [New York Times, October 27, 1870 p. 4]. Magnetic disturbances were recorded at the Melbourne Observatory in Australia, and found to occur at the same time as magnetic disturbances in northern observatories. [Harpers Weekly, 1871, issue 09/16].

November 9, 1871 – *"Brilliant Display of Aurora Borealis"*. One of the most striking and brilliant displays of the aurora borealis ever witnessed in this section occurred last evening and attracted great attention from those who were fortunate in being abroad at the time. About 6 o'clock the sky to the westward became suddenly illuminated by a red light, causing all who saw it in the City to believe that a terrible fire had broken out in Jersey City. Groups gathered in the streets and watched the illumination, speculating on the nature of the material that could produce so vivid a reflection. For over ten minutes the light grew in intensity and size, until it appeared as if the entire city on the other side of the Hudson was in flames. Then the theory of a fire was dispelled by a rapid shifting of the light to the north-east, and the aurora was fully developed in its full grandeur... The aurora was remarkably beautiful in Boston, and a brilliant display occurred at Hartford Connecticut, continuing at the later place all night." [New York Times, November 10, 1871, p. 1]

February 4-5, 1872 - An aurora was seen from Paris with rays that stretched to the zenith. It was not reported to be as spectacular as the 'blood red' aurora seen in October 1870. "It was a splendid show rising just over the Arc de l'Etoile and in broad crimson rays stretching up to the very zenith. Nearly everyone fancied that it was the reflection of a vast incendiary beyond Les Invalides or toward Autruil, and it was only when reporters who had hastily left their bocks to run off after an item, began to return that we knew the character of this grand display. It was not equal however to the grand aurora of 1870 and conversation turned upon recollections of the three October nights when the sky was blood-red from horizon to horizon."

February 29, 1872 [New York Times, p. 2]. It was also observed from Havana, Cuba and was widely interpreted to be an omen of the end of the world. [New York Times, February 17, 1872, p. 2]. Seen in Bombay, Sydney, Cape of Good Hope, Tobago, Cuba and Paris. Magnetic observatories in Bombay and Greenwich recorded a spectacular change in Earth's magnetism that exceeded 960-gammas; nearly as powerful as the 1,700-gamma change seen in Bombay during the Carrington-Hodgson flare, making it the 6th most powerful magnetic disturbance recorded between 1857 and 1938. From Havanna a New York Times news correspondent wrote "The number of men and women who read the destruction of the world in this sign was large and the latter took great care to go to church and pray that the calamity might be averted."

Aurora borealis March 1, 1872. (Credit: E. Trouvelot, New York Public Library)

August 18, 1872 - *It has recently been held that the aurora borealis is intimately connected with peculiarities in the solar photosphere. Whenever the aurora is frequent and brilliant, the black patches commonly called solar spots are numerous and active in their changes. This fact would seem to establish a connection between these two phenomena. Within the past month we have had the most remarkable auroral display that has occurred within the memory of the present generation, while it is a matter of general remark*

THE AURORA BOREALIS.

A BRILLIANT DISPLAY LAST NIGHT.

THE HEAVENS ABLAZE FOR MORE THAN TWO HOURS—EFFECT ON THE TELE-GRAPH WIRES—MESSAGES SENT BY THE AURORAL CURRENT TO PHILADELPHIA, ALBANY, MONTREAL, BOSTON, AND OTHER POINTS.

Such a brilliant display of northern lights as was seen in this City last night has not been witnessed for several years. There was no aurora at all last year, and not since the wonderful exhibition of the Spring of 1869 has anything so sublime been observed as the play last night of the fire that

that within the last twenty years the aurora, which was formerly rarely visible in this latitude, has become yearly more and more frequent. If, therefore, there is the connection between the sun and the aurora which is now believed to exist, we are warranted in concluding that unusual solar disturbances have recently been, and now are in progress.

It has moreover been fully established that the aurora is electrical in its character. During the week that has just passed not only has the aurora been almost nightly visible, but throughout the country we have had a series of thunder-storms unparalleled in the grandeur of their electrical features..." [New York Times, August 18, 1872, p. 4].

Long before photography was perfected to capture solar details, astronomers like Samuel Langley at the Allegheny Observatory would spend patient hours drawing sunspot details taking advantage of every moment when atmospheric turbulence was minimal. Here is a sunspot seen on December 23, 1873. (Credit: Robert Stawell Ball 'The Story of the Heavens', 1900)

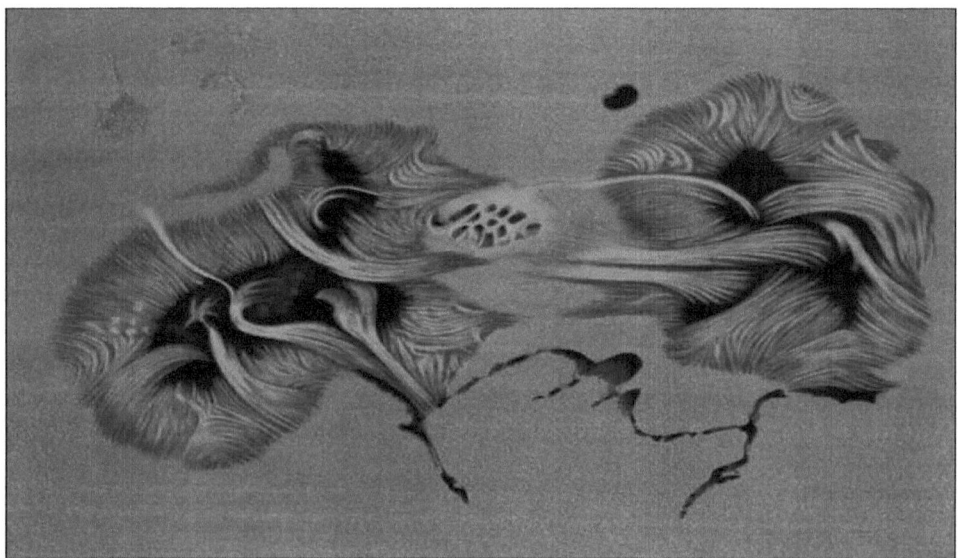

Sunspot drawing June 17, 1875 by Étienne Léopold Trouvelot, a talented French artist with a penchant for astronomy. (Credit: Trouvelot Astronomical Drawings Manual. 1882)

May 28, 1877 - It was observed as an arch that moved halfway to the zenith from New York City. Following a dramatic exhibition of curtains, waves and shooting rays extending to zenith, the display faded after 10:00 PM. Telegraph lines were affected from Boston, Baltimore, Philadelphia and Washington DC. *"The effect on the telegraph wires over a large area of country was to disturb them notably. At the Western Union Telegraph Office in this City all the quadruplex instruments were paralyzed and it was necessary in the case of a number of circuits to make new adjustments to overcome the disturbing influence. Experiments were made from the moment the current began to affect the wires, and the galvanometer was employed to note the movement of the current from north to south and from south back to north. The delicate needle of the galvanometer obeyed every 'shaking of the banners' as Mr. Dolan the electrician in charge termed the movement of the electric lights... From 9 o'clock until midnight the wires were last night so thoroughly charged that with the batteries disconnected it was still possible to communicate with many distant points."* [New York Times, May 29, 1877, p. 5].

Northern Lights over Calgary April 2003 (Credit: Wikipedia/Qyd)

The Chicago Daily Tribune [May 29, 1877 p. 2] reported that

'...the atmosphere was so strongly impregnated with electricity that communication was kept up for some time with New York, Boston and Montreal over the Western Union telegraph wires without the use of a battery. In Boston it was the most brilliant aurora seen in many years." In the next day's edition *[May 30, p. 10]* the aurora was seen in LaSalle, Illinois and Watertown, Wisconsin. *"A remarkably brilliant display of aurora borealis attracted the attention of many of the people in LaSalle last night. Successive waves of silvery light rolled up from the north and massed in luminous clouds near the zenith. The northwestern horizon displayed a brilliant crimson color, and a belt of silvery light flashed and wreathed half the expanse of the heavens from east to west.*

In Watertown, while Charles Krebs stood outdoors admiring the aurora borealis, the money-drawer was taken from his saloon, and all the cash it contained, to the amount between $3 and $4 stolen. The drawer, with some papers, was found in a tree a short distance away. Three persons were arrested on suspicion, but were discharged as nothing connected with the theft could be proven against them. One of them, however, named Top Rogers, was held for vagrancy and sentenced to the Jefferson Jail."

August 12, 1880 - Shortly after 9:00 AM, telegraph lines in Hartford, Connecticut began to show disturbances. With the battery removed, messages could still be sent and received from Boston. By 11:00 AM the wires were working as normal. No aurora were seen at night. "The cause of the phenomenon witnessed is not known, but is supposed by some electricians to have some connection with the aurora borealis, and if the disturbance had occurred in the night, it would probably have been accompanied by a brilliant display of Northern lights..." [New York Times, August 15, 1880, p. 8]

TOO MUCH ELECTRICITY.

THE EFFECT OF THE RECENT SEVERE MAGNETIC STORM.

Telegraphic Business Greatly Impeded in New York—Cable and Domestic Dispatches Delayed Several Hours —An Electrician's Views.

NEW YORK, September 13.—Last night, in various parts of this section of the country the sky presented a most striking appearance, so much so that astronomers had plenty of opportunity for making observations. This

September 13, 1881 – Too Much Electricity – The effect of the recent severe magnetic storm. *"Last night in various parts of this section of the country the sky presented a most striking appearance, so much so that astronomers had plenty of opportunity for making observations...The aurora was so strong that it lasted all night and was continued this morning, the consequence being that telegraphic communication by the eastern land wires was greatly impeded and the regular cable dispatches were consequently delayed several hours...A magnetic storm is always associated with a visible phenomenon known as aurora borealis, and generally believed to be the aurora of the Southern [magnetic] Polar regions [located in the Arctic circle]. The wires of the Western Union and other telegraph companies were greatly impeded for a time but as the day wore on the disturbances became less and business is not greatly retarded...Mr. George Hamilton, associate electrician of the Western Union Telegraph Company said to a reporter, '...These magnetic storms are always accompanied by aurora and earth currents, which are known to us as currents of electricity which traverse the surface of the globe, a portion of which is caught up by the telegraph wires and frequently occasions a serious disturbance in the communications...In respect to the bond which connects sunspots with magnetic disturbances, no positive conjecture has yet been formed, but the fact is eminently suggestive and only brings us at once into the presence of some great cosmical (sic) bond different from gravitation, and at the same time adding a new interest and mystery to these perplexing phenomena...The most important magnetic disturbances I remember here took place in 1867 and then in 1873 or74. The first of these fitful agitations of the needle was noticed in Berlin in as far back as 1807..."*[The Washington Post, September 14, 1881, p. 1]

April 16-17, 1882- New York City was bathed in a blood-red glow from a bright, 2:45 AM aurora, and continued until dawn. *"... While it was at its brightest, a paper might have been read in the open streets. The City was bathed in the blood-red light which streamed down from the firmament and*

painted the silent, deserted streets and structures. The effect was beautiful in the extreme, and entranced those who saw it." Astronomer Henry Draper is interviewed about the current scientific understanding of aurora. "The idea that the aurora was a destroying angel spreading pestilence over the earth had long since been given up, but the science of aurora was not yet entirely clear. The manifestations occurred at from 100 to 150 miles above the earth where the air was very rare. By running an electric current within an almost exhausted vessel the same effect might be produced."

In Chicago, a corona was observed at zenith. Telegraph lines between Chicago and Milwaukee, St. Paul and Omaha were 'worked' without batteries. Observations were also reported from Cincinnati, Davenport, and Cleveland. In Cleveland, a priest and other people with 'nervous disorders' were reported to have been effected. [New York Times, April 18, 1882]. Other reports were cited from Poughkeepsie, Baltimore, New London and Hartford Connecticut [New York Times, April 17, 1882, p. 5]. There were probably two events involved, one on each consecutive night.

The 1882 Transit of Venus Storm

November 18, 1882 - It produced a compass bearing deflection of nearly 2 degrees, All telegraphic transactions east of the Mississippi River and north of Washington D.C came to a halt. The Chicago stock market was severely affected all day. A large sunspot was then seen covering an area of more than three thousand millions of square miles. Simultaneously with the appearance of the spot, magnetic disturbances at the observatory in Greenwich increased in frequency and violence, other symptoms were noticed throughout the length of the British Isles. Telegraphic communication was greatly interfered with. The signal bells on many of the railway lines were rung, and some of the operators received shocks from their instruments. Lastly, on November 17, a superb aurora was witnessed, the culminating feature of which was the appearance, at about six o'clock in the evening, of a mysterious beam of greenish light, in shape something like a cigar, and many degrees in length, which rose in the east and crossed the sky at a pace much quicker than but nearly as even as that of sun, moon, or stars, till it set in the west two minutes after its rising. The daily press was burdened with accounts of widespread magnetic disturbance, in some places

A STORM OF ELECTRICITY

TELEGRAPH WIRES USELESS FOR SEVERAL HOURS.

ONE OF THE MOST SEVERE DISTURBANCES FOR MANY YEARS, EXTENDING EVEN TO EUROPE—TELEPHONE WIRES ALSO OB-STRUCTED—BUSINESS DELAYED A GOOD PART OF THE DAY.

Yesterday's storm was accompanied by a more serious electrical disturbance than has been known for years. It very seriously affected the workings of the telegraph lines both on the land and in the sea, and for three hours—from 9 A. M. until noon—telegraph business east of the Mississippi and north of Washington was at a stand-still.

telegraphic communication was suspended. In Milwaukee the carbons in the electric lamps were lighted, rendered incandescent by currents of electricity flowing on the wires. At other locations, switchboards in telegraph offices were set on fire and sending keys were melted, while electric balls were seen hovering on the telegraph in Nebraska.

A spectacular magnetic storm ensued when a large sunspot with an area of 4,700 millionths of the solar hemisphere, passed the meridian. It produced a compass bearing deflection of nearly 2 degrees, and a change in earth's horizontal and vertical magnetic fields exceeding 1060 gammas. The telegraph wires were useless for hours and business activity was interrupted for the better part of a day. All telegraphic transactions east of the Mississippi River and north of Washington D.C came to a halt.

Telephone lines of the Metropolitan Telephone Company refused to work until after 2:00PM. "...*people who attempted to use the telephones heard a buzzing, ringing noise rather than any well-defined sound while attempting communication, and occasional words only could be distinguished.*" Disruptions were also reported on the cables to Cuba and Mexico. The Chicago stock market was severely affected all day, There was an electric storm which downed the wires and left members of the Board of Trade largely to the devices of their own heads." [The New York Times, November 18, 1882, p. 1]

Scientific interest in this storm was considerable. For only the second time this century, numerous observations were collected within the limited pages of a major scientific journal. There were many (25) reports in Nature. Connection was made in most reports to 'a naked-eye sunspot, the largest seen at Greenwich, with spectroscopic information provided. Prof. C. Smyth, the Astronomer Royal for Scotland notes that Simon Newcomb in his 'Popular Astronomy' book suggested that sunspots and aurora are coincidental. The transit of Venus followed a few weeks later.

Rare sketch from the Boston Daily Globe, February 11, 1892 of the solar surface to show the large sunspot group responsible for the geomagnetic storms during this period.

February 13, 1892 - This was described as the 'most wonderful exposition ever seen on American soil' [LA Times, Feb 14, p. 1], and stretched from Iowa to the Atlantic coast. Was seen in Cleveland, Louisville, Detroit and Milwaukee, but not in St. Louis, Kansas City or Memphis. It was thought to be a fire by citizens of Plainfield NY. Students and townspeople in Princeton NJ watched it and regarded as a 'calamity' by some citizens. [New York Times, February 14, 1892, p.1].

July 16, 1892 - Nature once more paints the aurora borealis – but while esthetic souls admire the beautiful shimmering bars of light and color, telegraph operator says bad words, for electrical display interferes with the working of the wires…All the afternoon and early evening telegraphers whose instruments would not work properly knew that something was wrong up above, and it remained for the electrical display to demonstrate that it was…Between the hours of 10 am and 3 pm, while the sky was cloudless and the sun shining brightly, a tremendous current of electricity said to be due to the aurora borealis, charged the air to such an extent that it was impossible for the telegraph companies to use the long circuits on their wires while it lasted…The officials said it was absolutely impossible to get the use of the wires at Pittsburg, New York, Omaha and St. Louis."[Chicago Daily, July 17, p. 1] [The Washington Post, July 17, p. 3]

HER CANVAS THE SKY.

NATURE ONCE MORE PAINTS THE AURORA BOREALIS.

But While Esthetic Souls Admire the Beautiful Shimmering Bars of Light and Color, the Telegraph Operator Says Bad Words, for the Electrical Display Interferes with the Working of the Wires—During the Day Long Connections Are Shut Off.

Those citizens of Chicago whose immediate sky was not undercast with smoke were treated to a beautiful display of the aurora borealis a little after 9 o'clock last night. All the afternoon and early evening telegraphers whose instruments would not work properly knew that something was wrong up above, and

July 16, 1893 Brilliant display of aurora borealis seen at Chicago - Telegraph wires all over the Northwest were badly affected by the electrical disturbance. The Western Union experienced the most difficulty with its wires from Chicago to Omaha. [Chicago Daily Tribune, July 17, p. 1]

September 10, 1898 - The telegraph lines in Chicago were disabled by a 'daytime aurora borealis'. The effects were seen on telegraphs in Omaha, Tennessee, Washington. The shocks produced 280 volts on the lines. "A Western Union official said, we have had trouble with the Aurora Borealis before and it has been noticed that the weather was always as it is today. The shocks registered a force of 280 volts upon some of our lines. The trouble is widespread. Strange to say, the telephone lines were not affected. Forecaster Cox said "The Aurora was very strong and if the day had not been cloudy it could easily have been seen".[New York Times, September 10, 1898, p. 1]

A common red-green aurora curtain (Courtesy: Wikipedia/ Brocken Inaglory)

Electrical freaks of nature interfere with the telegraph; wires all over the United States east of the Rocky Mountains made useless by strange ground currents. Phenomenon accompanied by high barometric pressure – Had the aurora occurred at night Chicago probably would have been treated to the most brilliant atmospheric display in years...At 3 pm, there was not a single telegraph wire working east of Pittsburg or Buffalo. Atlanta and Augusta reported their lines north to Washington out of service. After 4 o'clock the magnetic display ceased altogether and the wires worked as well as before. [Chicago Daily Tribune, September 10, p.7]

Aurora have also been detected on Saturn and Jupiter by the Cassini and Galileo spacecraft. As for Earth, energetic particles from the sun disturb the planetary magnetic field, which then accelerates trapped particles into the polar regions of these planets. (Credit: NASA/ESA)

5 Storms of the 20th Century

The 1903 Geneva Street Car Storm

AURORA STOPS THE KEYS.

**ELECTRICAL FREAKS OF NATURE IN-
TERFERE WITH THE TELEGRAPH.**

**Wires All Over the United States East
of the Rocky Mountains Made Useless by Strange Ground Currents—
Phenomena Accompanied by High
Barometric Pressure—Their Occurrence at This Time of the Year Considered Remarkable.**

Telegraph wires over all the United States
east of the Rocky Mountains refused to work
yesterday between 11 a. m. and 4 p. m., during which time an electrical disturbance
similar to that sometimes occasioned by the
aurora borealis was in progress. The effect
on the wires was much like that of an elec-

November 1, 1903 - Telegraph systems of Western Union were affected from 2:00AM this afternoon. This was identified as most severe storm since 1888 according to Chief Electrician for WU. Transatlantic cables were also affected. Marconni Wireless Telegraph Company said they were not affected at all. [New York Times, November 2, 1903, p. 7]. Magnetic storm seen in France, Switzerland but not Austria, Italy or Denmark. But Swiss streetcars were disabled when power went out. Aurora seen in Ireland and Scotland. Sir Oliver Lodge and Norman Lockyer attributed the event to sunspots, which were also blamed for unusual wet weather. [New York Times, November 2, 1903 p.1]. Aurora borealis puts telegraph companies out of business – *"…The disturbance lasted eight hours. At its climax there were 675 volts of electricity – enough to kill a man – in the wires without batteries attached. It was the worst electrical disturbance in thirty-five years, said Chief Operator J. E. Pettit of the Postal Telegraph Company, 'At times there were no workable wires in any direction and the cable service was seriously affected both on the Atlantic and Pacific".* [New York Times, November 1, 1903, p. 3]. Spots on the sun cause trouble: Strange phenomenon in France and Switzerland [New York Times, November 2, 1903 p.7]. Electric waves sweep the world: Mysterious pulsations of energy rain down from the North Pole. [New York Times, November 1, 1903 p.8]. Electrical disturbances due to spots on the sun [New York Times, November 8, 1903 p. A45]. Aurora borealis paralyzes wires [New York Times, November 1, 1903 p. 2]

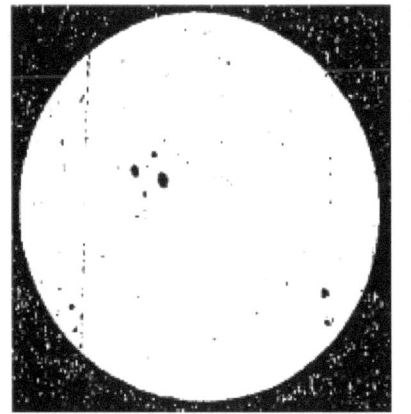

Sketch of sun showing sun spots for November 4, 1903 (Chicago Daily Tribune, November 8, 1903, p. A45)

The magnetic storm from this four-day event caused a 3-degree compass change in the bearing for magnetic North, and a magnetic field change similar to the 1882 storm. The New York Times reported telegraph disturbances in France that had cut them off from the rest of the world from 9:00 AM to 4:00 PM. Submarine cable traffic into, or out of, France was also disturbed. Because the most intense phase of this magnetic storm happened in broad daylight, there were no sightings of aurora to tip off French citizens about what was going on above their heads to cause all the electrical ruckus.

The front page New York Times story 'Northern Lights Display' described these difficulties in terms of the aurora which were by this time so brilliantly seen in the US. Here, both the aurora and telegraphic interruptions began at the same time. At the offices of the Marconi Wireless Telegraph Company, W. H. Bentley said of the disturbance that it was one with which the company had never had to contend with as yet." It had not, apparently, had any effect on wireless telegraphy even in the more northern latitudes. A second article on the front page "Made the Wires Deadly" describes how ...at its climax there were 675 volts of electricity, enough to kill a man, in the wires without a battery attached [in the lines entering Chicago] Both Atlantic and Pacific cable service was affected in the most severe storm anyone could remember in the last 30 years.

This storm was notable, not only for its by-now familiar communications mischief, but for a new kind of disturbance that no one had anticipated. France was generally identified as the epicenter for the worse effects of this daylight storm, but in Switzerland they were experiencing collateral effects at the same time, and these led to an event that no one had ever seen before.

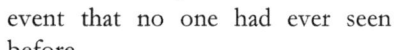

Electric street cars like the ones shown above in Ottawa, Canada were common in all major cities in the United States and Europe ca 1900. (Credit: Ontario Canada Transpo Collection:Flickr)

The Swiss telephone and telegraph services ceased suddenly and remained inoperative for half an hour.   But in Geneva, all the electrical street cars were brought to a complete standstill. The unexpected cessation of the electrical current supplying the cars was a mystery to the engineers. Despite their efforts, they were not able to re-start the cars until after the magnetic storm had passed. "Telephone and Street Car Service Suspended in Switzerland for Half an Hour Saturday – Scientists attribute the magnetic disturbances of yesterday to sunspots. The worst effects of the phenomena appear to have been experienced in France. Berlin was not affected and apparently neither Austria, Italy nor Denmark suffered. In Switzerland, however, there occurred a most strange phenomena. The telephone service ceased suddenly and remained suspended for half an hour while the telegraphs were rendered unintelligible and useless. In Geneva all electrical street cars were brought to a sudden standstill, and the unexpected cessation of the electric current caused consternation at the generating works where all efforts to discover the cause were fruitless. The Meteorological Office reports a magnetic storm, accompanied by aurora borealis, in several parts of Ireland and Scotland. The telegraphic disturbance was one of the most extraordinary on record. Late this morning before it had been settled where the disturbances originated, although Siberia was strongly suspected slight seismic disturbances were reported from the Riviera. [New York Times, November 2, p. 7]

ELECTRIC PHENOMENA IN PARTS OF EUROPE.

Telephone and Street Car Services Suspended in Switzerland for Half an Hour on Saturday.

LONDON, Nov. 1.—Scientists attribute the magnetic disturbance of yesterday to sunspots. The worst effects of the phenomena appear to have been experienced in France. Berlin was not affected, and apparently neither Austria, Italy, nor Denmark suffered.

In Switzerland, however, there occurred a most strange phenomena. The telephone service ceased suddenly and remained suspended for half an hour, while the telegraphs were rendered unintelligible and useless. In Geneva all the electrical street cars were brought to a sudden standstill, and the unexpected cessation of the electric current caused consternation at the generating works, where all efforts to discover the cause were fruitless.

"Two severe electrical disturbances, one following the other so closely that in some parts of the country they seemed as one, have practically held up telephone and telegraph business in the United States, Canada and Mexico for more than twenty four hours, and experts declare that similar disturbances may recur at frequent intervals for a week… In Chicago telegraph and long distance telephone lines were seriously interfered with, and wireless telegraphy was almost impossible. From points in the north and western part of the United States came reports of a brilliant display of the northern lights, and even in Chicago above the smoke of the city a faint red glow hung in the northern sky. In Salt Lake City the aurora was mistaken for a fire north of the city Saturday morning, and in Duluth Minnesota it took the form of huge waving columns apparently blown by the wind…"Theoretically such a disturbance as that of yesterday, if intensified, might exterminate all life on the face of the earth, but there is little likelihood of this. Had the currents been more continuous and of a slightly higher voltage they would have been perceptible to the human body" said Prof. Cox of the Weather Bureau…[Chicago Daily, Nov 1, p.8]

March 2, 1905 - Electrical disturbances on telegraph lines from Chicago to Sioux City affected telegraph lines. Aurora predicted for evening. "An electrical disturbance in connection with the aurora borealis is predicted for today on a large scale on account of spots on the sun, has affected telegraph wires from Chicago west as far as Sioux City, Iowa. Prof Cox of the Weather Bureau said that on such occasions magnetic storms were usually created. "It is an electro-magnetic storm and the difference in the working of the wires indicates that we are getting a good deal more electricity than ordinary' he said. The lights are expected to be visible tomorrow New York Times, March 3, 1905, p. 1].

MAGNETIC STORM GRIPS THE EARTH

Electric Currents, Said to Be Due to Sun Spots, Cripple Telegraph Service.

WIRES ARE "STAGNANT."

Similar Disturbances Every 11 Years, Accompanied by Aurora Borealis.

September 25, 1909 - Telegraph lines throughout US were affected. Some wires carried 500 volts of electricity and lit incandescent 'resistance lamps' in telegraph circuits. [New York Times, September 26, 1909, p. 12]. Aurora borealis stops telegraph [New York Times, September 26, 1909, B4]. Magnetic storm grips the earth. [New York Times, September 26, 1909, p.6] Aurora

borealis stops telegraph communication [New York Times, September 26, 1909, p. I7].
Aurora upsets wires:Mysterious electrical storm sweeps two continents [New York Times,
September 26, 1909 p.3].

The Advent of Radio

While a silent battle was being waged between telegraphists and aurora, Alexander Graham
Bell, in 1871, uttered the first telephonic sentence in his laboratory "Mr. Watson, come
here. I want you". In less than a year, the first toll-free, long-distance phone call was placed
by Watson and Bell, between Cambridgeport and Boston, using borrowed telegraph lines.
Meanwhile, as if to celebrate this event, the Great Aurora of February 4, 1872 colored the
skies. Again, reports could be found in the newspapers and science journals of powerful
voltages induced upon telegraph lines. During the November 17, 1882 Great Aurora, the
telephone lines of the Metropolitan Telephone Company refused to work until after
2:00PM. Disruptions were also reported on the cables to Cuba and Mexico. The Chicago
stock market was severely affected all day. By the time impacts were identified, it was
already far too late to rethink the deployment of the technology. The impact that solar
storms had on telegraph, telephone, and power lines was truly unexpected. By the time the
famous September 1859 storm had lashed the Earth, telegraphy had already become a
transcontinental reality, displacing the Pony Express with 30,000 miles of line strung up on

trees and poles. Telephony was born 19 years
later, but it took another 11 years for its
vulnerability to be tested during the November
18, 1882 solar storm.

Telephone wires from many different companies on
poles in downtown New York City ca 1889.

By 1901 there were over 855,000
telephones in service in the 'Bell Telephone
System'. It seemed as though the telephone
industry had taken the country, and the world,
by storm. Everyone wanted their own private
line, and the only limiting factor in the spread
of this technology was how quickly the Bell
Telephone Company could cut down trees to
make telephone poles, and wire your city block
or town into the growing national circuitry.

No sooner had some considerable
money been spent on wiring the world for telegraph and telephone, but a still newer
technology appeared in full bloom from literally out of nowhere. Guillermo 'William'
Marconi in 1895 tinkered together the first spark gap radio wave transmitter and receiver in
the garden of his father's estate. Instead of transmitting and receiving electrons flowing in a
wire, it was the 'wireless' emission and reception of electromagnetic radiation that carried
the messages. By 1905, there were over 100 wireless telegraph transmitters in the US with

transmission ranges of 500 miles. There were also some seven million telephones in service on the same wires that once carried telegraphic messages.

British Post Office engineers inspect Guglielmo Marconi's wireless telegraphy (radio) equipment, during a demonstration on Flat Holm island, 13 May 1897. This was the world's first demonstration of the transmission of radio signals over open sea, between Lavernock Point and Flat Holm Island, a distance of 3 miles.

September 25, 1909 - This magnetic storm lasted only about 10 hours and affected a large part of the world. Its disturbances caused a 3 ½ degree compass change, and a spectacular 1,500-gamma change in Earth's horizontal magnetic field – the largest recorded since the August 28, 1859 storm. Of all the severe storms between 1857 and 1938, it ranked as the third most severe recorded by instruments at the Potsdam Magnetic Observatory. This storm was most severe in America, England and Europe. Telegraphic communication throughout the British Isles were stopped, and there were up to 6 hour delays in sending or receiving messages from France. Telegraph lines as far south as Uruguay were affected, and telegraph service in New York was badly crippled. Electrical surges on some lines that exceeded 500 Volts were also reported in telegraph offices in New York City. Brilliant sparks leapt across the gaps when the telegraph keys were opened. The current flowing in the wires also lighted the incandescent resistance lamps in the telegraph circuit. The disturbance started at 7:00 AM and became progressively worse as the day unfolded.

New Yorkers, who usually are among the first to spot aurora covering their skies even from Central Park, were not able to see the aurora despite the clear skies, but this circumstance had little to do with the brightness of the aurora itself, had it been present. On this particular night, beginning at 8:00 PM to midnight – prime aurora viewing hours, a massive celebration was taking place in honor of Hudson and Fulton. Over a million people were out on the streets, walking by the riverside, or driving along Riverside Drive to see the Naval ships all decked out in electric lights. This celebration involved lighting up the entire city and the Hudson River with powerful searchlight beams and fireworks displays. No fewer than 20, fifty-million candle power searchlights beamed colored light into the sky in a display that was 'simply unparalleled…and the greatest display ever seen in the world'. No aurora could have competed with the vast wattage being pumped into the skies above New York on this night.

William Marconi discussed the interference that the storm had caused over a large part of the world. US and English telegraph systems . *"I can't help being a little glad that the telegraph companies have had this object lesson...Wireless is affected by certain things which do not hinder the ordinary lines, but in this matter we have the advantage"* [New York Times September 27, 1909 (page 9)] With the expansion of wireless technology, the trap had now been set to prove Marconi's boast rather premature. In a world that was to suffer through two world wars and other military conflicts, the consequences could not have been more upsetting.

The 1915 Wireless Radio Storm

May 24, 1915 - Germans have trouble communicating out of country due to severe radio interference. Intelligence gathering operations greatly hampered, and had to rely on use of British resources...and censors.[New York Times, May 25, 1915, p.3]. Aurora borealis halts wireless to Germany [New York Times, May 25, 1915, p.15].

AURORA BOREALIS HALTS WIRELESS TO GERMANY.

[BY A. P. NIGHT WIRE.]

NEW YORK, May 24.—Wireless communication between the United States and Germany has been severely handicapped and probably will continue so until about July 1 by the static conditions prevailing in the North Atlantic at this time of the year.

The electrical activities of the aurora borealis, accompanied by electrical storms over the wireless routes, are the cause of the difficulty. It is explained, and may be expected to continue several weeks. This means that until normal conditions again prevail uncensored communication between Germany and the outside world will be limited

The news agencies meantime have been obliged to depend upon London for regular transmission of the daily German War Office statement.

A 'wireless outage' in Northern Europe caused by an aurora left Allied intelligence agencies in England and the United States unable to gather information about Germany for weeks leading up to World War I. The normally direct channel between Germany and the US was unavailable due to severe static, and instead the communication had to proceed through Great Britain, where British censors were hard at work on the messages. Germany was virtually isolated from the rest of the word via direct transmissions unless the British censors allowed the messages to go over the Allied-controlled cables instead. Thanks to the auroral static disturbances, Germany had to fall back on wireless channels to get her messages out to the rest of the world without going through English censors and Allied-controlled channels. The German-owned Atlantic cable was cut at the

beginning of the war. They were able to work around this loss by using their long-wave station in Sayville Long Island which they owned through their Atlantic Communication Company to relay their uncensored messages from the German War Office across the world to their diplomatic representative in neutral countries. This station, however, was cut off as well by the Allies.

June 17, 1915 - North wires unaffected: Only Eastern telegraph circuits affected by aurora [New York Times, June 18, 1915, p. 13]. Aurora holds up telegraph – The bewildering beauty of the aurora borealis which lighted the whole of the northwestern United States and Canada Wednesday night was replaced Thursday with extensive reports of interruptions to telegraphic service caused by its electric currents. Until 9 o'clock today the Western Union Telegraph Company reported interrupted service across northern Idaho, Montana and the Dakotas where night service was virtually suspended. [New York Times, June 20, 1815, p.12]. A brilliant aurora was observed early today at the Lowell Observatory in Flagstaff Arizona. The northern horizon was peculiarly bright and ruddy at midnight…The electrical disturbances caused by the aurora borealis reported this morning [June 17] from various western states had a serious effect also on cable and telegraph lines in the northeastern part of the United States and eastern Canada. For several hours during the early morning cable communication via the Newfoundland cables of the Western Union was all but paralyzed. [Los Angeles Times, June 18, p. I3]

August 26, 1916 - Wire service lamed by aurora borealis – The disturbance according to the officials was first observed about 7 pm. By midnight it was said the service of the telegraph companies east of Chicago and as far west as Minneapolis was only about 50 percent normal. Southern circuits were not much affected.[The Washington Post, August 27, 1916, p.A1].

August 9, 1917 - Aurora borealis monkeys with telegraph lines [Chicago Daily Tribune, August 9, 1917, p. 1]. Earth currents knock out wires – The disturbance began soon after 11 o'clock affecting wires between New York and Atlanta. Later it was felt on all wires between New York, Buffalo and Pittsburgh and even as far west as Chicago. Lines to Boston and to the cable station at North Sydney Nova Scotia ceased to operate. Officials of the Western Union Telegraph Company said that fully 90 percent of the wire facilities in the country east of Chicago had been temporarily put out of commission. [The Washington Post, August 9, 1917, p. 2]

The 1918 London Bombing Aurora

March 9, 1918- Telegraph lines from New York to Buffalo were disrupted. Motors providing electricity for the telegraph wires were acting strangely. No one could understand how 'atmospheric electricity' could affect motors. [New York Times, March 9, 1918, p.9]. In London, the auroral light aided German bombers in seeing terrain over southern England. [New York Times, March 9, 1918, p. 3]. Strange light in sky watched by crowds [New York Times, March 8, 1918, p. 11]. Ojibway indians say celestial apparition portends great events [New York Times, March 9, 1918, p. 3]. Two officers chased aurora borealis thinking it fire. For the first time in the present generation the aurora borealis was visible in the northern sky from Tampa tonight. There was a vivid red glow as if a great fire was raging. One report had it that Dade City, a town forty miles to the north, was a-fire, but Dade City reported the same glow to the north. [The Atlanta Constitution, March 8, 1918, p.1].Aurora on spree of color paints the sky red [Chicago Daily Tribune, March 8, 1918, p. p. 13]. An aurora borealis glows in northern sky startles capitol [The Washington Post, March 8, 1918, p.1]. Experts deny London raid due to aurora borealis [The Washington Post, March 11,

GERMAN AIR RAIDERS KILL 11 IN LONDON

46 Wounded by Bombs Dropped from Airplanes in Thursday Night's Attack.

AURORA AIDS THE GERMANS

Sky Brighter Than in Moonlight— Only Two of Seven or Eight Machines Reach the City.

1918, p.3].

The aurora also helped light up the English landscape enough that seven or eight German bombers were able to stage an air raid over England and bombed parts of London around 11:45 PM. Just before the raiders were spotted flying over Kent in southern England, a bright aurora filled the sky with more light than a full moon. "There was a remarkable display of the aurora borealis last night and it is believed by many that this furnished conditions under which the air raiders could work more effectively than under a clear star-lit sky. Watchers on the Kent coast said that just before they heard the raiders approaching, the whole northern sky became illuminated in bands of red and white light which shone over the sea with far more powerful effect than the full moon." [New York Times, March 9, p.3]

Another aurora that took the skies on March 9, 1918 did more than disturb telegraph lines running from New York to Buffalo and Halifax. The motors that furnished electricity to run the Western Union telegraph wires began to act strangely during the aurora. It wasn't known how the 'electricity in the air' could operate to interfere with the electrical apparatus and to change the 'quality of the current' generated. At the radio station at the Brooklyn Navy Yard, reporters questioned the Navy about the 'freakish electrical show' and its impact upon wireless communication. Their questions were refused answers on the grounds that 'all information of any character regarding wireless operation had to be furnished through official channels in Washington' [New York Times, March 9, p. 9]

February 3, 1919 - Red Artilliary shells Petrograde Seized in revolt [The New York Times, February 3, 1919, p.1].

October 2, 1919 - Aurora borealis cripples wires [The New York Times, October 3, 1919 p.3]. Aurora borealis makes splash in our midst – Long before the calls came, the telegraph and long distance phone companies knew the sun's storm of electrons was playing havoc with its service. "The trouble started on our lines early in the evening' said the night superintendent of the long distance 'It developed at New York then hit Pittsburgh and Buffalo and finally reached this district. Now we are catching it out Omaha way and up towards Minneapolis'. [Chicago Daily Tribune, October 2, 1919, p. 1].

August 11, 1919 - Surplus of atmospheric electricity is blamed as the cause of telegraph disruptions along Atlantic seaboard as far south as Georgia. 'Scientists' quoted as saying that aurora have nothing to do with telegraph line problem. [New York Times, August 12, 1919, p. 8]. Wire system halted by aurora borealis – The electrical phenomenon put out of commission thousands of miles of wire and made its influence felt as far south as Kansas City, according to reports to the Western Union Telegraph Company.[The Los Angeles Times, August 12, 1919, p. I9].

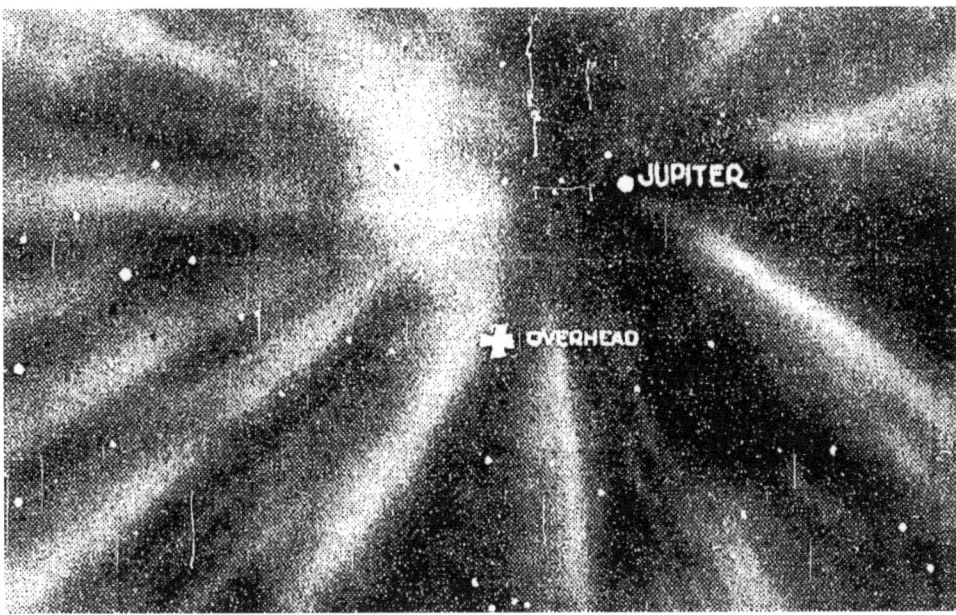

Sketch of aurora over Chicago, March 23, 1920. (Credit: Chicago Tribune artist S.F. Maxwell)

March 22, 1920 - American Telephone and Telegraph telegraph service disrupted - Atlantic cable traffic reduced from 30 messages per day to one. Aurora seen over New York City. "In the Bronx many women and children to whom the meaning of the flares and flashed had not been explained, ran into their homes and pulled down shades in order to shut out the mysterious glow" [New York Times, March 23, 1920, p. 13]. Aurora borealis seen in Atlanta – Quite a crew of staid Atlantans were astonished by the phenomenon for they didn't know whence, what or why. Darktown darkened perceptibly and its popular prediction that prayer meetings will rival mullet suppers for popularity with the dusky tribe for a few days. The hardened old skeptics that scoff like regular humans breath snatched at the contention that somebody was advertising a new brand of chewing gum or hair net. The quite universal and bibulous class believed it had found another adverse test for the lumber yard attributes of alcohol." [The Atlanta Constitution, March 23, 1920, p. 1]. Jazzy aurora snarls wires, dazzles city [The Chicago Daily Tribune, March 23, 1920, p. 1]. The aurora borealis as Chicago saw it [Chicago Daily Tribune, March 23, 1920, p.5].

The 1921 New York Railroad Storm

May 13, 1921 - The prelude to this particular storm began with a major sunspot sighted on the limb of the sun vast enough to be seen with the naked eye through smoked glass. The spot was 94,000 miles long and 21,000 miles wide and by May 14th was near the center of the sun in prime location to unleash an earth-directed flare. The 3-degree magnetic bearing change among the five worst events recorded ended all communications traffic from the

Atlantic Coast to the Mississippi. At 7:04 AM on May 15, the entire signal and switching system of the New York Central Railroad below 125th street was put out of operation, followed by a fire in the control tower at 57th Street and Park Avenue. No one had ever heard of such a thing having happened during the course of an auroral display. The cause of the outage was later ascribed to a 'ground current' that had invaded the electrical system. Railroad officials formally assigned blame for a fire destroyed the Central New England Railroad station, to the aurora. Telegraph Operator Hatch said that he was actually driven away from his telegraph instrument by a flame that enveloped his switchboard and ignited the entire building at a loss of $6,000. Overseas, in Sweden a telephone station was 'burned out', and the storm interfered with telephone, telegraph and cable traffic over most of Europe. Aurora were visible in the Eastern United States, with additional reports from Pasadena California where the aurora reached zenith. [Newspaper Archive]

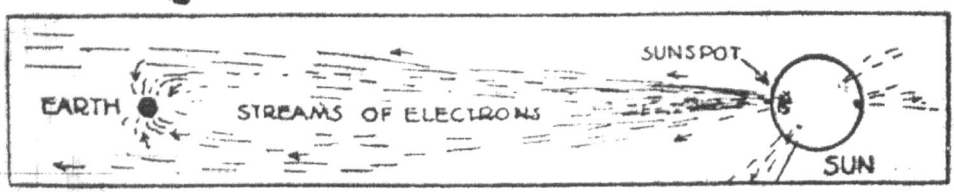

Illustration from the May 15, 1921 Chicago Daily Tribune (page 12) showing how sunspots create the aurora borealis.

The prelude to this storm began with a major sunspot sighted on the limb of the sun by Dr. George Peters at the Naval Observatory on May 9th. By May 13th it was now vast enough to be seen with the naked eye through smoked glass. The spot was 94,000 miles long and 21,000 miles wide and by May 14th was near the center of the sun in prime location to unleash an earth-directed flare. What was odd about this very large spot was that it came near sunspot minimum conditions, which arrived in 1923, It also sat nearly completely across the equator of the sun. That location probably explained why, within a day, the spot divided itself like a splitting cell, into two spots that now took up their locations each in its own hemisphere of the sun. At about this same time, a powerful series of magnetic storms commenced and lasted for four days.[H.W. Newton, 'The Face of the Sun' p. 96]. The 3-degree magnetic bearing change and 1000-gamma magnetic field changes that resulted during this 3-day storm were among the five worst events recorded at the magnetic observatory in Potsdam between 1857 and 1938.

It was described as 'an aurora borealis unparalleled in dimensions in the memory of telegraph wire chiefs', and by midnight it had ended all communications traffic from the Atlantic Coast to the Mississippi. The last major event of its kind was considered to be the aurora on Sept 25, 1909. The storm began to interfere with telegraph and telephone companies on Friday night. It would take two hours to send messages that normally take 15 minutes to clear. In New York City, theatergoers 'on the way home who changed to look at the sky last night, rubbed their eyes, decided to sign the pledge and then looked once more. It was still there – the aurora borealis". Until after midnight, crowds stood on Broadway, not to chat about the latest stage productions, but to gape at the lights flashing over their heads, which covered the entire sky from the northern horizon to the zenith in curtains of shimmering crimson and violet light.

The storm produced large earth currents in the Post Office telegraph systems in England, Scotland and Ireland, and was given considerable attention in newspapers such as the London Times of May 19. The sunspot cycle reached its peak in 1917 and this storm was associated with an unusually large sunspot group that appeared on May 8.

By 10:00 PM May 15, Washington DC was cut off telegraphically from the rest of the United States. What was even more exciting, and potentially disastrous, was that the Chiefs in charge of wire traffic at the Chesapeake and Potomac Telephone headquarters were seeing their lines carrying more than 1000 volts of electricity.   President Newcomb Carlton of the Western Union Telegraph Company described how the magnetic disturbances had 'blown out fuses, injured electrical apparatus and done other things which had never been caused by any ground and ocean current known in the past'. The company would probably have to send ships to drag up the undersea cables to repair them. The electrical ocean currents had found the weakest spots in the cable insulation and caused severe damage. Apparently three of the Western Union trans-Atlantic cables were affected. By the next day, May 17, two had been place back in operation, but the third cable was damaged somewhere near Valentia Island.

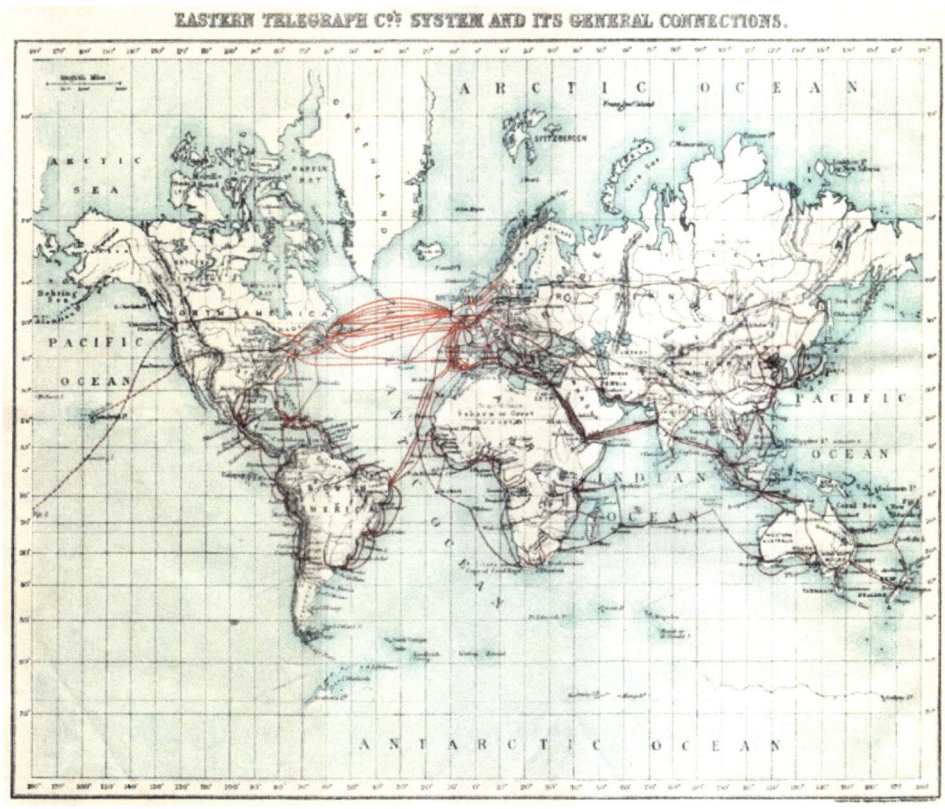

Map of the Eastern Telegraph Company undersea telegraph cables ca 1901 (Credit:Wikipedia)

Finding a problem in a cable involves carefully measuring the resistance of the cable between the station and the 'break'. By knowing the precise resistance of the cable in ohms per meter, engineers can figure out to the nearest mile where the break would be.

The challenge is then to send a special repair ship to that location and establish the ship's location exactly over the break. The inch-thick cable then has to be carefully snagged and hoisted several thousand feet to the surface. The strain on a cable can be considerable and repair is always a very risky business, however for a line that is not working anyway, it is a gamble most are willing to take. In this instance, however, it would cost $200,000 to fish the magnetic storm-damaged section out of the sea and repair it by splicing in a new section. More problems were soon to follow.

At 7:04 AM on May 15, the entire signal and switching system of the New York Central Railroad below 125th street was put out of operation, followed by a fire in the control tower at 57th Street and Park Avenue. No one had ever heard of such a thing having happened during the course of an auroral display. The cause of the outage was later ascribed to a 'ground current' that had invaded the electrical system. The repercussion was that the automatic signaling and switching devices as well as the railroad's telephone and telegraph system stopped immediately. Fumes came from the insulation of the electrical wired in the control tower, and dense smoke followed. The Fire Department was called, but by the time they arrived, the dense clouds of noxious smoke from the burning insulation had invaded the homes of many Park Avenue residents blocks away. The failure of the electrical switching forced railroad engineers to literally turn the switches by hand so that some rail traffic could flow. Grand Central Station was thrown into pandemonium with thousands of delayed passengers waiting for the next available trains to take them to work. Miles Bronson, head of the electrical division of New York Central described how the insulation of a third-rail jumper had broken down and become connected with a water pipe, causing the short circuit. Normally, the current flow would have been carried south on other conductors, but some electricians believed that the magnetic disturbances had changed the normal current flows so that they went northward and set off the control tower disaster. The fact that the wire chiefs had detected 1000 volt surges on their lines caused Bronson to be open to the possibility that with those kinds of voltages loose in the ground, the New York Railroad outage could easily have been a consequence.

Meanwhile in Brewster New York, railroad officials formally assigned blame for a fire destroyed the Central New England Railroad station (see modern photo), to the aurora. Telegraph Operator Hatch said that he was actually driven away from his telegraph instrument by a flame that enveloped his switchboard and ignited the entire building at a loss of $6,000. Overseas, in Sweden a telephone station was 'burned out', and interfered with telephone, telegraph and cable traffic over most of Europe. The aurora were visible in the Eastern United States, with additional reports from Pasadena California where the aurora reached zenith.

By the time of this aurora, there were 1.5 million radio receivers in use painstakingly assembled on kitchen tables and workbenches by legions of eager new customers. Only the advent of the PC computer would again find comparable numbers of 'hobbyists' likewise assembling their systems on the floor. In 1924, some 15 million telephones were in use on the Bell System, and the first transmission of pictures over a phone line became a reality.

The Bronx Sisters tuning their ca 1925 radio. (Credit: Bain News service photo via Library of Congress)

So what did people do with this new technology? Many people sure didn't use it very responsibly. Unlike the telephone or telegraph where the ends of the lines are geographically known, for wireless broadcasts, everyone is anonymous unless they choose to identify themselves. As Historian Edward Herron wrote in <u>Miracle of the Air Waves: A history of radio</u>.

"..[Amateurs] thrilled to calls for help from sinking ships...and were not above creating synthetic excitement...sending out false messages that caused international distress, confusion,

and waste of time and resources...Commercial stations depending on the dollar revenue from the dots and dashes, were constantly at war with the amateurs who rode ruthlessly into the same wavelengths, causing havoc with the commercial messages."

This forerunner to modern computer 'hacking' was the main reason why the US Government had to step in and put an end to the unruly amateur broadcasts in 1917. Once World War I had concluded in 1919, the embargo was lifted, and the pace of radio technology research exploded like champagne out of a bottle. The first commercial radio station, KDKA, owned by Westinghouse opened for business on November 2, 1920 to a hungry crowd of over 30,000 amateur wireless operators who had cobbled together their receivers as home hobbyists. Two years later, there were 1.5 million sets in use, and by the end of the decade there were radio sets in 7,500,000 homes. This phenomenon had taken eight years to escalate to this level, while telephone service took 37 years to reach the same number of homes. Today's stampede of people onto the Internet is only the most recent of many waves of colonization of new high-tech niches that have opened up during this century.

Radios in 1925 were assembled using the same approach as building cars. The work was mostly done by women with close-cropped hair. (Credit: Library of Congress)

Most of the broadcasting during the 1920's was done at long wavelengths, but by 1925 the Navy got involved with short-wave broadcasting because it could be received long distances with little interference, and could also be transmitted during the daytime, unlike the then-popular long wave transmissions. Wars are fought day and night, so there was tremendous pressure to push transmission technology to higher frequencies and shorter wavelengths where daytime 'bounce' was possible.

But wars are fought day and night so there was tremendous pressure to push transmission technology to higher frequencies and shorter wavelengths where daytime 'bounce' was possible as shown in the figure below. Ironically, the short-wave radio frequencies would drive communication into the very domain that made then a victim of solar interference. Now, whenever aurora dominated the sky, and the Sun was throwing out flares like electromagnetic thunderbolts, their impacts would appear in many different

guises, and across the entire spectrum of communications technology. By the time that sunspot cycle 17 began in 1933, 23 million homes (70% of total homes) had short-wave radio receivers, and Americans listened to nearly one billion hours each week of broadcasting. Television receivers operating in the newly conquered megacycle radio spectrum were already being field tested by several manufacturers and were expected to be available to the consumer within a few years. Short-wave interruptions were an increasingly common annoyance during daytime broadcasting, but their origins in distant solar flares were not recognized until 1937.

So long as radio communication was limited to nighttime long wave broadcasts, no one really considered the sun a nuisance. As the technology continued to develop and expand relentlessly, radio waves were eventually found to suffer from problems just as annoying as the lessons learned by telegraphists. On occasion, and without warning, daytime shortwave reception would fade out for many minutes at a time before recovering to its original daytime conditions along the particular 'circuit' being used. The reason for these costly radio problems had been investigated in 1925 by the National Bureau of Standards, and they concluded that the problem could be traced to solar activity.

March 8, 1924 – Aurora borealis is causing trouble for wire companies and radio operators. Later in the month with the approach of the vernal equinox the aurora causes trouble of no mean proportions because it is a phenomena the electrical engineers have not learned to cope with. The auroral disturbance comes about this time every year and returns again in the fall. Some trouble was experienced by the telephone and telegraph companies last night while there was considerable difficulty in transmission of code messages by radio. Entertainment programs, however, were not affected, it was said at the broadcasting station of the Chesapeake and Potomac Telephone Company. [The Washington Post, March 8, p.5]

January 17, 1926 – The aurora borealis or northern lights is blamed for recent erratic radio reception, particularly during the last two nights, by W.D. Terrell, chief radio inspector of the Department of Commerce. While the presence of the aurora borealis in sufficient strength to affect telegraph and telephone transmission has not been reported to the Weather Bureau, such electrical disturbances Mr. Terrell declared, might be present even though not seen. No remedy has been found for the interference. [New York Times, January 17, p.21]

The 1926 Stock Market Crash Storm

OBSERVATIONS SHOW AURORA HAS LITTLE EFFECT ON RADIO

Strange Currents Disturb Telegraph Lines— MacMillan Says Radio Is Not Hampered By Northern Lights in Arctic

January 26, 1926 - Electrical disturbances believed to have been caused by aurora borealis interfered with telegraphic communications in most of the United States at intervals yesterday…"The interference did not affect the long-distance telephone lines or the long-distance telegraph lines, which are metallic throughout. It did affect the grounded telegraph circuits and interference amounting to as much as 120 or 150 volts of foreign potential was discovered. As far as the Bell system is concerned, the interference affected only grounded telegraph lines throughout the country except southeast of Atlanta."[New York Times, January 27, p. 1] Observations show aurora has little effect on radio [The New York Times, January 31, 1926, p. XX15]. Aurora turns telegrapher [The Los Angeles Times, January 28, 1926, p.5].

Aurora borealis offers best alibi for the Bulls in stock market – "Aurora borealis was the best alibi of the bulls market today when the stock market turned up its toes for a midwinter nap, and the wire systems that spread like spider webs throughout the country to catch the unwary fly, suddenly ceased to function. "Have many out of town orders come in? said the senior at the end of the first hour to the wire manager. "Nothing doing", "Why?", "I can't get anything west of Hoboken", "What's the cause?", "Search me" said the manager. But the cause was much farther north than his own personality. And this was the explanation of the American Telephone and Telegraph Company 'Interference with the system's telegraph and telephone lines out of New York presumably by the aurora borealis began at 11:19 am and afterwards at intervals of from two to ten minutes up to noon. Interference amounted to 120 volts of foreign potential noted." [The Washington Post, January 27, 1926, p. 10]. Borealis cavorts on wires [The Los Angeles Times, January 27, 1926, p.6].

March 11, 1926 – Salzburg calls fire brigade to put out Aurora Borealis – It was the first time the aurora borealis had ever been seen there and so many alarms were turned in from all parts of Salzburg that the bewildered firemen thought the whole city was simultaneously in flames and helped to increase the panic by dashing about in all directions. Peasants in the outlying country fled, thinking judgment day had arrived.[New York Times, March 11, p. 1]

April 15, 1926 – Residents of Mineola, Garden City and other Long Island towns had an exceptionally fine view of the lights. The white streak extended from the north in a southerly direction. Reports from Northern New York and New England States said that the phenomenon was also visible there. Radio, telegraph and telephone communication was somewhat hampered by the electrically charged atmosphere [New York Times, April 15, p.2] Chicago has colorful night in glow of aurora borealis [Chicago Daily Tribune, April 15, p.6]

The 1926 Coolidge Speech Storm

October 15, 1926 – Magnetic storm swept over Northeastern US and Canada. Stock market transactions between London and New York were slowed down, but were completed by the afternoon. Auroral beams shot towards zenith over New York City. Wires were affected as far south as Dallas Texas and west to Denver [New York Times, October 16, 1926, p. 11]. Aurora faint in south and west [The New York Times, October 16, 1926, p.11]. Aurora interferes with wires and cables by surcharging earth with electricity [The New York Times, October 16, 1926, p. 11].

Aurora borealis hits Coolidge speech. Engineers explain failure of the radio – The aurora borealis was blamed today by scientists for the failure of President Coolidge's speech at the International Oratorial Contest last night to get out through the air to radio listeners. F.P. Guthrie, district manager of the Radio Corporation of America, said his engineers held the aurora borealis responsible. Tests were made on a telegraph line leased by the radio station at 7:30 o'clock, but when the radio operator took his post at the broadcasting instrument at 8:05 he found the signals from the Washington Auditorium so faint he could not send them out. The engineers looked for the cause but could find no mechanical trouble. After a twenty-minute silence, the Auditorium signals increased in volume and the remainder of the program could be broadcast. The aurora borealis yesterday affected all telegraph and telephone wires throughout the northern part of this continent. [The New York Times, October 17, 1926, p.3]. Aurora paralyzes wires – Slow transmission of market transactions between Wall Street and London was caused by the magnetic disturbance this morning, but by the time the height of the disturbance was reached this afternoon most of the trans-Atlantic market transactions had been completed. [The Los Angeles Times, October 16, 1926, p. 2].

February 25, 1927 - In Geneva, villagers asked fire department to extinguish the aurora. [New York Times, February 26, 1927, p. 15]

April 14, 1927 - Aurora borealis isolates West from Eastern Canada - The aurora borealis played havoc with wire communications from Winnipeg to Montreal today, and for a time practically isolated the West from Eastern Canada. The strength of the aurora lessened somewhat at noon and communications were re-established. The disturbance was the strongest in many years and effected an almost complete tie-up in telegraph communication with United States points south of here.[The New York Times, April 15, 1927 p. 23].

July 22, 1927 - Wire service affected - [The New York Times, July 22, 1927, p. 19].

October 13, 1927 - Freakish currents slow wire service [The New York Times, October 13, 1927, p. 19]. Electric disturbance upsets German radio – The greatest electrical disturbances in eleven years have been recorded by the Meteorological Magnetic Observatory at Potsdam. Sun spots recurring periodically are held responsible for them...The short wave radio service to northern countries is utterly broken down and open telegraph lines have been seriously affected. The entire day the northern sky was a hazy purple and tonight there is the glow of the aurora borealis. Radio service to the Equator and beyond is better than usual. This is explained by the area of greatest disturbances being around the magnetic pole. The long wave radio, however, has continued to function north though the signals are weaker than usual. [The New York Times, October 14, 1927, p. 22]. Wire lines twice hit by aurora – The leased wire network of the Associated Press was seriously affected and reports from the Western Union and Postal Telegraph companies indicate that their service was more or less crippled...Aurora borealis is a natural phenomenon which has puzzled scientists for years and had been the bane of telegraph operators since the invention of telegraphy. [Los Angeles Times, October 13,1927, p.3].

July 8, 1928 - Telegraph lines are tied up by aurora borealis – Telegraph messages sent into and out of Chicago last night were seriously delayed by earth currents thrown out by the aurora borealis...The tie-up affected wires all over the country. At New York, San Francisco and Atlanta, main trunk lines were only partly serviceable for several hours. [Chicago Daily Tribune, July 8, 1928, p. 2]. Aurora plays queer tricks [Los Angeles Times, July 8, 1928, p. 3].

The 1930 Short-wave Era Storms

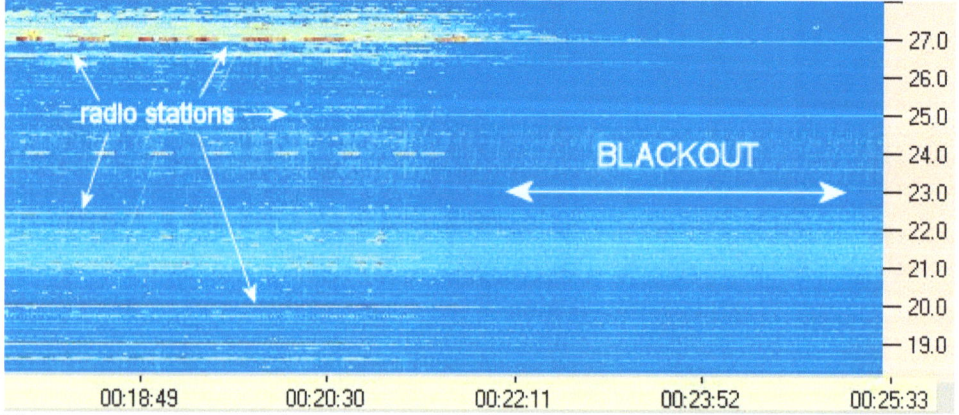

This spectrum graph, created by astronomer Dick Flagg shows that after 00:22 UT on April 24, 2014 when a solar flare erupted, the transmissions of all the short-wave stations were 'blacked out'. The vertical scale gives the station frequency in MegaHertz. (Credit: Dick Flagg, Windward Community College, Ohau; and http://www.astrowatch.net/2014/04/x-class-flare-erupts-from-sun-causes.html)

March 2, 1930 – Magnetic storms hovering over the Atlantic are bothering the radio engineers who are striving to pick up European programs for rebroadcasting in America. The transoceanic talk-bridge of the American Telephone and Telegraph Company, however, seems to be less affected than the National Broadcasting Company's short-wave channels. Last Sunday, Lady Astor spoke in London and was heard with remarkable clearness as her words crossed the sea on the American Telephone and Telegraph circuit for rebroadcast over WABSC's system. A half hour later the WEAF-WJZ network with WOR linked into the hook-up tried to rebroadcast from Holland, but the bombardment of the magnetic storms ruined the clarity. [New York Times, March 2, p.133]

May 30, 1932 - Borealis lights up night skies disrupts wires [Chicago Daily Tribune, May 30, 1932, p. 1].

September 9, 1933 - Midnight sky lighted up by aurora borealis – Telegraph companies reported that the electro-magnetic waves had caused an interruption of their service from the northwest but that there was no interference to the east and the south.[Chicago Daily Tribune, September 9, 1933, p. 1].

June 10, 1936 - Telegraph and short wave radio service interrupted by electrical disturbances – The American Telephone and Telegraph Company reported interference with the short-wave radio telephone but said the long-wave radio phone and the long-distance wires were unaffected. The disturbances had cleared up by 10 am. The difficulty was attributed to sun spots. [New York Times, June 20, 1936, p.19]

June 19, 1936 - Telegraphic transmission hit by aurora borealis [Chicago Daily Tribune, June 20, 1936, p.13].

February 28, 1937 - Spots on sun are blamed for freak reception of tiny waves [New York Times, February 28, 1937, p. 174].

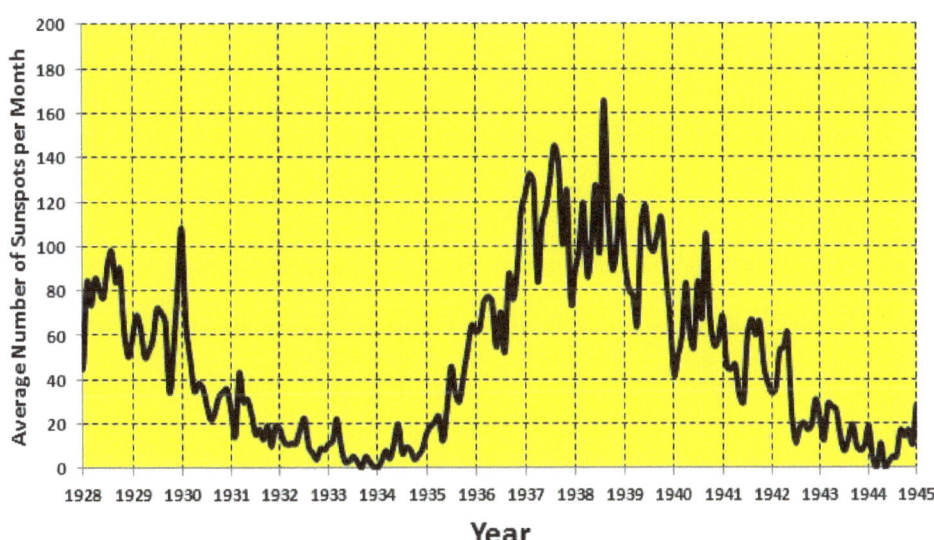

During this period, the average number of sunspots seen on the sun was approaching its maximum number for the current 11-year cycle: Number 17.

April 28, 1937 - "Worst magnetic storm in a hundred years' continued for past 4 days. Magnetic disturbances correlated with large sunspot. Clouds conditions prevented aurora from being seen over eastern US. From the prairies to the Maritimes and into the United

States, telegraph wires were gripped by the excess current radiated from the northern magnetic pole. Wires were grounded as operators in Canadian National and Canadian Pacific offices fought to maintain connections... An unusual feature was the appearance of the interference on the short-distance circuits in Ontario and Quebec and down to Halifax. [New York Times, April 29, 1937, p. 23]. Aurora borealis hits wire services - [The New York Times, April 29, 1937, p. 23].

August 3, 1937 - Brilliant aurora lights Niagara skies [The New York Times, August 3, 1937, p. 25].

The 1938 Lady of Fatima Storm

January 25, 1938 - The Great Aurora was seen over the whole of Europe and as far south as Southern Australia, Sicily, Portugal and across the Atlantic to Bermuda and Southern California. The Japanese invasion of China was the main news on the international front. All transatlantic radio communication was interrupted. Crowds in Vienna awaiting the eminent birth of Princess Juliana's baby cheered the aurora as a lucky omen. Fire department of Salzburg was called out to quench what residents thought was their town in flames. So many alarms were rang that the fire department dashed about in all directions, which only served to increase the level of panic among the citizens. This same impression of the aurora also struck Londoners during the January 1938 aurora who also thought their entire city was aflame. In England, a hook-and-ladder brigade was summoned to Windsor Castle to put out a nonexistent fire. [New York Times, January 30, p.43] In San Diego, forest officials in the town of Descanso about 40 miles east, were routed out of bed on January 22 to respond to reports of 'great fire in the back country'. After making the trip to check things out, all they discovered upon arrival was the crimson aurora borealis in the northern sky, last seen in these areas on February 1888. [New York Times, January 23, p.29] In Bermuda, many people thought that a ship was on fire at sea. Steamship agents took the precaution of checking with wireless stations to learn if there had been any S.O.S calls. Meanwhile, in Scotland, many of the more superstitious people living in the lowlands 'shook their heads and declared the northern lights always spelled ill-omen for Scotland.

The phenomenon also had some side effects. It was responsible for delaying express trains on the Manchester to Sheffield line after electrical disturbance hit the signaling apparatus. Numerous false impressions were aroused among Cannock Chase people. One person thought there was a big fire at a local colliery and phoned the fire brigade. In some quarters it was said the world was coming to an end. Short wave radio sets were interfered with and the teletype system at the local office of the Western Union was started up by the phenomenon. Astronomers in New England said the lights differed from previous auroral displays with such intensity and color and direction of the beams. The immense arches of crimson light with shifting areas of green and blue, radiated from a brilliant Auroral Crown near the zenith instead of appearing as usual in parallel lines. It was also considered to be one of the Fatima Prophesies by Roman Catholics worldwide.

This Great Aurora was seen over the whole of Europe and as far south as Sicily, and in California and Bermuda. The Japanese invasion of China was the main news on the international front, but the New York Times on January 26 ran an article on page 25 describing how the Britons were dazzled by the biggest display they had seen in 50 years. All trans-Atlantic radio communication was interrupted. Crowds in Vienna awaiting the imminent birth of Princess Juliana's baby cheered the aurora as a lucky omen. Princess Juliana and Prince Bernhard gave birth to Princess Beatrix a few days later on January 31.

They had not seen a similar aurora since March 9, 1926 when the fire department of Salzburg was called out to quench what residents thought was their town in flames. So many alarms were rang that the fire department dashed about in all directions, which only served to increase the level of panic among the citizens. This same impression of the aurora also struck Londoners during the January 1938 aurora who also thought their entire city was aflame. The aurora was seen in Italy and Spain. Bermuda observers saw it very clearly with dark red streamers. Commercial telegraph lines in Toronto, Montreal and Ottawa were severely affected.

Aurora Borealis Startles Europe; People Flee in Fear, Call Firemen

Britons Thought Windsor Castle Ablaze— Scots See Ill Omen—Snow-Clad Swiss Alps Glow—Short-Wave Radio Halts

"The phenomenon was also the cause of delay to express trains on the L.N.F.R. Manchester-Sheffeld line. At 7:48 PM, the signaling apparatus in both the parallel Woodhead Tunnels was found to be out of order. The working of the trains through the tunnels was stopped. An official said that the failure was apparently due to the electrical disturbances caused by the Aurora Borealis." (London Times, 1-27-1938, p. 2)

The "aurora borealis" is a luminescent meteor, a phenomenon that frequently happens in areas close to the North Pole and which can also be seen in rather exceptional circumstances in regions of Central Europe. So the aurora borealis that could quite clearly be seen from the Pyrenees, and even from the top of the Tibidabo hill in Barcelona, on the 25th of January 1938, was an absolutely unusual occurrence. It was in fact a unique experience. There are no known accounts of any other event of that kind at such meridional latitudes. Furthermore, the phenomenon took place in the midst of war, thus causing terrible confusion and shock among the soldiers who were fighting on the Aragonese front. (Barcelona Journal,1999)

The 1938 Aurora Borealis – Looking North (Dalskairth, Scotland) watercolor by Scottish Vincent Robert Stewart Ramsey Balfour-Browne (1881 – 1963); now hanging at the Dumfries Museum in Scotland.

"Dazzled by a natural light show: West Midlanders thrilled in January to the Northern Lights which hung over Britain for two hours. Astronomers said the brilliant display in the sky was due to "the outsize in sunspots" which had made their way to the other side of the sun. The phenomenon also had some side effects. It was responsible for delaying express trains on the Manchester to Sheffield line after electrical disturbance hit the signaling apparatus. The Midlands got the best display of the aurora borealis where the sky was spasmodically lit up by strange lights. One Express & Star reader reported them over Sedgley and said the sky was filled with "bright red, luminous, feathery clouds." Another observer at Wolverhampton's Goldthorn Hill, said there was a deep, red glow in the sky like the reflection of the glare from a blast furnace fire. Numerous false impressions were aroused among Cannock Chase people. One person thought there was a big fire at a local colliery and phoned the fire brigade. In some quarters it was said the world was coming to an end" [Westmidlands.com, 2014]

April 16, 1938 – The most powerful magnetic storm experienced since August-September 1859 was detected at the Potsdam Magnetic Observatory. At its peak, it upset compass bearings by an astonishing 5 ½ degrees and caused a 1900-gamma change in earth's magnetic field – nearly 3%. It ranked as the second most powerful storm since 1859, and there were some similarities to the 1859 storm. An initial storm event arrived at 06:00 UT on August 16th and lasted two hours, followed by a second, and much longer storm event that continued for nearly a full day afterwards.

August 20, 1939 – Best sunspot show since 1870 is fading – For the first time short-wave radio fadeouts were linked directly to eruptions occurring near sunspots. These fadeouts ranged from fifteen minutes to half an hour. Telegraph and telephone service was disrupted at intervals by magnetic storms on earth caused by solar spots and aurora borealis

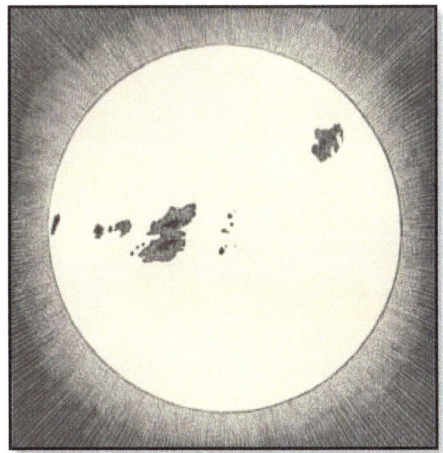

displays were seen as far south as the Mexican border. Ships compasses behaved erratically at times. [New York Times, August 20, p. 32]

Sketch of sunspots seen on August 30, 1939.

October 14, 1939 Electrical disturbances described popularly ass aurora borealis interrupted telegraph communications in Ontario, New York State and Western Canada today. Wire traffic was interrupted for varying intervals as a result. [New York Times, October 15, p. 9]

The 1940 Easter Sunday Storm

Sun-Spot Tornado Disrupts Cables, Phones and Telegraph for 5 Hours

Electrical Disturbance Plays Havoc With the Short-Wave, 1,000,000 Easter Messages and Police and Press Teletypes

March 25, 1940 - On Easter Sunday calls to grandma by millions of people were halted between 10:00 AM and 4:00 PM creating pandemonium at nearly all Western Union offices. [New York Times, March 25, 1940, p. 1]. A telephone cable between Fargo North Dakota and Winnipeg was found with its wires fused together, presumably from the voltage surges. Consolidated Edison of New York also reported 1,500 volt dips in three electrical generators in New York City located in Brooklyn and the Bronx. In Bangor Maine, lightning arresters were burned out as well. The New York Times noted that United Press reported earth currents at 400 Volts in Boston, 450 in Milwaukee, and more than 750 Volts near St. Louis. All told, the Associated Press's entire investment of 185,000 miles of leased wires were put out of service. Practically every long-distance telegraph or telephone office in the country was doing repair work in what was considered one of the worst such events in history. AT&T land lines had been badly disrupted by 600 volt surges on wires designed for 48 volts. In the Atlantic Cable between Scotland and Newfoundland, voltages up to 2,600 volts were recorded during the storm. Coast Guard radio stations were blocked, although compasses were not affected. Excessive voltage in the Boston and Keene telegraph lines 'blew fuses'. In several instances fuses were 'blown' and vacuum tubes ran the risk of damage due to these influences. Earth counts toll of sun-spot storm [The New York Times, March 26, p. 18]. Nature's prank upsets the air [The New York Times, March 31, 1940, p. I34]. Sunspot ties up radio and wire service [Los Angeles Times, March 25, 1940, p.1]. Invisible sunspot storm over half of world disrupts cable, telegraph and radio [The Washington Post, March 25, p.1].

All shortwave traffic and news broadcasts between the United States and Europe was blacked out since 11:00 AM, although the long-wave channels that do not involve the ionosphere, were hardly affected. Postal Telegraph officials reported 200 to 400 volt surges in their service lines, with over 50% of all their lines affected in one way or another.

On CBS, the early broadcast of the Pope's Easter Message from the Vatican got through OK, however at NBC, they had to cancel their popular 3:30 PM European news broadcast. The Executive Curator of the Hayden Planetarium, William Barton, went on a nation-wide radio hookup to explain what was happening and causing all the trouble. A second storm occurred on Sunday, March 24 which also upset telephone and telegraph services world-wide. America was completely cut-off from the rest of the world for 2 hours. "If this disturbance had come on a weekday when the radio traffic was heaviest, it would have been almost catastrophic" In some places it really was catastrophic.

A telephone cable between Fargo North Dakota and Winnipeg was found with its wires fused together, presumably from the voltage surges. Consolidated Edison of New York also reported 1,500 volt dips in three electrical generators in New York City located in Brooklyn and the Bronx. In Bangor Maine, lightning arresters were burned out as well. The New York Times noted that United Press reported earth currents at 400 Volts in Boston, 450 in Milwaukee, and more than 750 Volts near St. Louis. The possibility that long-distance phone line technicians might be badly shocked from the high voltages on these lines, prompted American Telephone and Telegraph to stop work on these lines until after the magnetic storm had passed. The impact of these large voltages was to burn out fuses, carbons and grounded wires on the Western Union across their entire system.

All told, the Associated Press's entire investment of 185,000 miles of leased wires were put out of service. Practically every long-distance telegraph or telephone office in the country was doing repair work in what was considered one of the worst such events in history. Only the fact that many of Western Unions trunk lines are protected by electrical circuits that were not grounded, protected them from a complete breakdown of service.

AT&T land lines had been badly disrupted by 600 volt surges on wires designed for 48 volts. In the Atlantic Cable between Scotland and Newfoundland, voltages up to 2,600 volts were recorded during the storm. WBZ radio reported that its broadcasts were disrupted and faded erratically even over the local Boston service area. Coast Guard radio stations were blocked, although compasses were not affected. Excessive voltage in the Boston and Keene telegraph lines 'blew fuses'. In several instances fuses were 'blown' and vacuum tubes ran the risk of damage due to these influences.

At 1700 hours, all fuses for current (0.5 amp) on all lines burnt through. Smoke came from burning insulation material in coupling racks. The voltage between Earth and the cables could not be measured as no meters for such high voltages were available. A 230-volt [vacuum tube] coupled between Earth and ground after the most intense phase of the perturbation lighted up very intensely and had to be uncoupled at once in order to avoid damage."

The New York Times also reported that it was impossible to estimate the financial costs of the previous day's storm in terms of delayed services, and damaged equipment because of its wide-spread nature.

Electrical power companies had supplied a widening net of consumers since the first 225-home, lighting system was installed in 1882 by Thomas Edison. The stealthy effects of geomagnetic disturbances took a very long time to reach a threshold where their impact could actually be registered. A few extra amperes from celestial sources went entirely unnoticed for a great many years. The watershed event came with the March 24, 1940 solar storm, which caused a spectacular disruption of electrical service in New England, New York, Pennsylvania, Minnesota, Quebec and Ontario. By then, it was entirely too late to do much about the problem. Power grids had already become extensive and commonplace.

April 3, 1940 - Aurora borealis flares are visible in the city – Both Western Union and Postal Telegraph said there were signs of a magnetic disturbance on their long lines, but that the disturbance was not enough to interrupt the sending of messages as during the recent sunspot activity. The New York Telephone Company reported that there was no sign of the magnetic effects on their lines. [The New York Times, April 3, 1940, p. 21].

January 18, 1941 - Communications disturbed slightly by aurora borealis [Chicago Daily Tribune, January 18, 1941, p.10].

The 1941 Playoffs Storm

September 18, 1941 - This storm had the misfortune of occurring during a home game of the Brooklyn Dodgers and the Pittsburgh Pirates. During the day, baseball fans expected to hear the entire 4:00 PM broadcast on station WUR by Red Barber. With the game tied at 0-0, the station became inaudible for 15 minutes. When it resumed, the Pirates had piled up not just one, but FOUR runs. Within minutes, thousands of Brooklyn fans had pounded the radio station, demanding an explanation for the 'technical difficulties', only to receive the unsatisfactory answer that the sun was to blame. The effects of the 'sunspots' also appeared in the by-now usual problems with transatlantic short-wave communication to Europe, which was out for most of the day. But the sun wasn't quite finished with New York City after the baseball game 'mishap'. [New York Times, September 19, 1941, p. 25].

Aurora Borealis Gives City a Show As Sun Spots Disorganize Radio

Display of Northern Lights Unparalleled in Years Seen From Buffalo to Virginia— Dodger Fans See Red as Broadcast Fails

The next day, Sept 19, at 11:45 AM station WAAT was broadcasting some recorded music by Bing Crosby, when a conversation between two men was injected into the music, and rather clearly at that. There was absolutely nothing that the engineers could do about it. The merging of 'programs' was allowed to continue because although it was annoying, the voices were not deemed strong enough to warrant stopping the radio broadcast entirely. Then after a few minutes the mysterious voices vanished, and Crosby's music came through loud and clear again. Had this been the only problem, listeners would not have noticed. But a few minutes after the men's voices ceased, a new pair of voices emerged from Crosby's singing. This time the conversation was far from mild. The topic of their conversation, overheard by millions of listeners, was a blind date, and the discussion was rather 'spicy', by all accounts.

Sun Spots Add Some Radio Spice, Sneaking Private Phone Calls on Air

Girls' Discussion of Blind Date Mysteriously Gets on Sedate Program as Magnetic Storm Continues—More Auroras Unlikely

Although the cross-talk lasted only a few minutes, it was enough to cause listeners to again pound the stations switchboard demanding to know why such dialog had been permitted during a family listening time. [New York Times, September 20, 1941, p. 19]. Cosmic brush paints Chicago sky with light [Chicago Daily Tribune, September 19, 1941, p.1]. An exhibition in the sky [Chicago Daily Tribune, September 22, 1941, p.10]. Aurora borealis disrupts radio and telegraph service [Los Angeles Times, September 19, 1941, p.1]. Aurora borealis helps R.A.F. in bomb raids [Los Angeles Times, September 20, 1941, p. 7]. Northern light show visits south [The Washington Post, September 19, 1941, p. 1]. Nazi cities hit as northern lights illuminate raiders goals [The Washington Post, September 21, 1941, p. 1].

The World War II Storms

During World War II, when short-wave messages relayed critical military information, not just civilian business transactions or letters to distant relatives, there were major radio blackouts on September 18, 1941, Overseas radio blackout lasts 18 hours – CBS/NBC shortwave disruption. Additional short wave blackouts during World War II happened on July 6, 1941. July 8, 1943 - A severe shortwave outage blanketed Europe and Moscow for 18 hours, and a particularly troublesome one a month later on September 3 interfered with the radio transmission of the Allied invasion of Italy.

July 6, 1941 - Sunspots silence shortwave radio [New York Times, July 6, 1941 p. 20] Sunspots and radio [New York Times, July 8, 1941 p. 18] Shortwave channels to Europe are affected [New York Times, July 8, 1941 p. 10] Outburst on sun made radio fade [New York Times, July 13, 1941 p.23]. Magnetic storm upsets communication lines [Los Angeles Times, July 6, 1941, p. 18]. Aurora borealis slows war news [The Washington Post, July 6, 1941, p. 4].

Aurora Borealis Helps R.A.F. in Bomb Raids

One of Heaviest Night Attacks in Weeks Sweeps Nazi Points From Norway to France

LONDON, Sept. 20. (AP)—The R.A.F. in one of the biggest offensives in weeks, begun by the light of the aurora borealis and continued throughout the daylight hours, pounded bitterly defended objectives from Norway to France today.

June 27, 1942 - Bremen burns after biggest raid [The Washington Post, June 27, 1942, p. 1].

September 4, 1943 - Sunspots hamper radio, delay news of invasion [New York Times, September 4, 1943 p. 2]

The invasion of Italy …..BBC Radio: "British troops have landed on the Italian mainland four years to the day after war was declared on Germany. Their arrival in the "toe" of Italy follows two days of heavy bombardment by warships and Flying Fortresses of railways and communication lines. After crossing the Straits of Messina from Sicily to mainland Italy, British and Canadian troops of the 8th Army met little resistance at the port of Reggio di Calabria.

To meet the challenges of conducting a war in the face of frequent radio blackouts, in 1943 the US Army Signal Corps devised an ingenious solution. They would install six long-wave transmitters in Newfoundland, Greenland, Iceland and Great Britain that could maintain a continuous radio link even during short-wave blackouts. Radio telegraph and teletype communications with the front lines could be operated 24-hours a day. Some of the towers were 180 feet tall and had to be erected in the face of intense winds, and to withstand 160 mph winds. [New York Times, December 30, p. 8]

October 15, 1944 – Aurora lights sky over wide Midwest area – Accompanying a magnetic storm, the aurora last night did not fail to cause some disturbances. Press Wireless reported it had lost contact with Belgium, Paris and London, and trans-Atlantic cable lines were slowed and messages garbled. [Chicago Daily Tribune, October 15, p. 1]

December 17, 1944 - Spots on sun disrupt phones – Electrical disturbances attributed by some engineers to sunspots or the aurora borealis, played hob with communication lines throughout the country today. A spokesman for the American Telephone and Telegraph Company said the disturbances hampered or interrupted service for short periods from coast to coast, with the greatest amount of difficulty in the West. AT&T said it expects them to continue for 36 hours. [Los Angeles Times, December 17, 1944 p. 5].

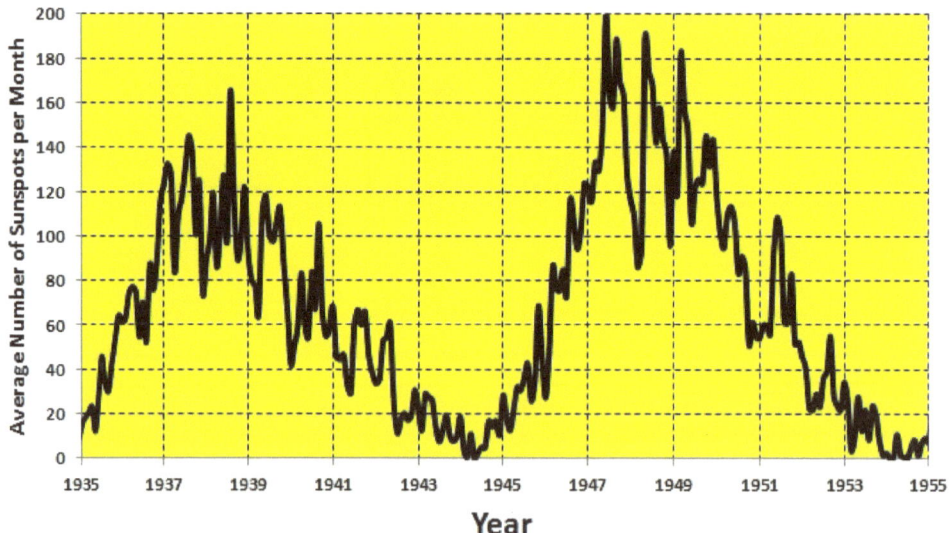

Although most of the military activity during World War II occurred as solar activity was drawing to a minimum in 1944, with barely a pause, solar activity began its predicted rise once again as the 1940s drew to a close, reaching its maximum between 1947 and 1949.

February 3, 1946 - Magnetic storms predicted to 'sweep earth' for next 12 days. It has begun with radio reception problems. Bombay, Lisbon, Cairo, and Singapore report telegraph disturbances. The last serious disruption was cited as March, 1940. [New York Times, February 3, 1946]. Green curtains, sheets and rays seen over New York City. CBS reported that 'sunspots' had caused an almost complete blackout of radio signals for second day. [New York Times, February 8, 1946, p. 18]

March 24, 1946 - New York and Canada report seeing aurora and describe it as one of the most spectacular in a decade. Red arches and streamers swept the heavens in the northwest. [New York Times, March 24, 1946 p. 13] Long-range radio disruptions caused air traffic delay over the Atlantic. – Long-range radio communications have been so seriously disrupted by the aurora borealis in the last few days that trans-Atlantic planes have in many cases been seriously delayed, according to airline reports yesterday. Thirteen planes operated by major transocean airlines were held up during the day. Six Europe-bound planes were stalled at Gander, Newfoundland, and seven westbound ones at Shannon, Eire. The Civil Aeronautics Administration's communications bureau attributed the condition to a spectacular display of the lights that began last Friday. The ionic disturbances, a reaction from storms on the sun, have prevented proper reflection of radio waves. [New York Times, March 27, 1946 p. 13]

April 8, 1946 - Electrified gases knock out radio [New York Times, April 8, 1946 p.17]

July 26, 1946 - Aurora seen over New York, Philadelphia and identified with sunspots now on sun. [New York Times, July 27, 1946, p. 23], Chicagoans see sky alight with auroral

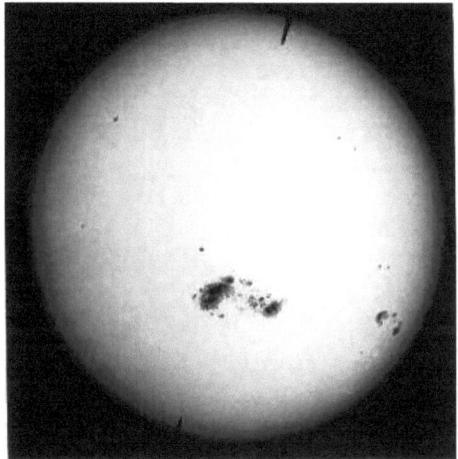

display – A sunspot 'curtain' across world air lanes today snarled international radio communications for the second consecutive day and interfered with domestic telegraph transmissions. Tonight, northern lights appeared over many cities. [Chicago Daily Tribune, July 27, 1946. p. 5].

April 8, 1947 large sunspot group (Credit: Mt. Wilson Observatory/ Carnegie Institution)

Now at the height of its 11-year cycle of activity, the years 1947-1948 led to a veritable hailstorm of solar flares as sunspot cycle 18 reached its crescendo. The sunspots of 1947 actually created a new high-level mark for solar activity that had not been seen before. The average for May was over 200! This year was also unique in that the largest sunspot ever to have been see on the sun was spotted in April. This enormous spot covered 6 billion square miles, and was so hardy that it was observed for four rotations of the sun in February, March, April and May. Most spots never make it around even one time before they fall apart. Curiously, with all the sunspot activity in April and May, this was not a remarkable year for major geomagnetic storms and auroral activity.

March 8, 1947 - World radio traffic garbled by sunspots – World-wide radio communications were all but blacked out yesterday by sunspots and magnetic disturbances, according to representatives of broadcasting stations and commercial message circuits. Radiotelephone service to Europe was reported out of order because of interference last night, while reception from South America and the Pacific was said to range from poor to good. The National Broadcasting Company was able to pick up only one of six remote-control news broadcasts scheduled for 8 A.M. yesterday. In one instance the program was from London, but the reception was described as poor. The network reported no word from Jerusalem or Berlin. Paris was said to have been scratchy to the point of being unintelligible, while a Buenos Aires offering faded just as badly. Moscow, usually picked up through London, could not be tuned in. [New York Times, March 9 1947, p. 11]

July 19, 1947 - Sunspots delay planes; Solar storms cause fadeouts in radios at Shannon – Sunspots that caused radio fadeouts and thus delayed transatlantic air traffic for many hours early yesterday morning were reported by the United Press yesterday from Shannon Airport, Eire. It was explained at La Guardia Field that planes on the international routes are not scheduled to arrive at a certain hour but either in the morning, afternoon or evening. However, some that ordinarily arrive in the morning did not get in until afternoon. The time lost, it was said, was on the Shannon to Gander leg of the route and not between Newfoundland and La Guardia Field. A spokesman for Pan American Airways reported interference with radio contact to Gander. Catherine Barry, assistant curator of the Hayden Planetarium said that a very large sunspot had been observed in the center of the sun. The storm on the face of the sun, she commented, might cause interference in radios and electronic devices, and make a display of the northern lights visible.[New York Times, July 18, 1947 p.15]

August 23, 1947 - Radio to Orient fades out – Atmospheric disturbances dropped an impenetrable curtain tonight over nearly all communications between the Orient and the West Coast of the United States. The interference, attributed both to sunspots and to equinoctial storms, increased in such intensity that even radio signals were erratic between San Francisco and New York. [New York Times, August 23, 1947 p. 4]

March 16, 1948 - Sunspots black out radio around world – A magnetic storm in the upper atmosphere has disrupted radio communications all over the world, the National Bureau of Standards reported today. The radio blackout began last midnight and is expected to continue three or four days. Bureau scientists said the disturbances are the worst in a year, and are particularly severe across the North Atlantic. They said the condition is caused by a heavy outburst of sunspots. Short wave communication was all but obliterated yesterday by sunspots, it was reported by the four major networks. Scheduled pickups from various parts of the world were said to have been unintelligible for broadcast. Only two signals, both from Paris, were said to have come through successfully by spokesmen for the Columbia and American chains. A short while later, the National Broadcasting Company was unable to raise the French capital.[New York Times, March 16, 1948, p. 36]

January 27, 1949 - Radio disturbance due to second largest sun spot – The second largest single spot ever noted on the sun was named yesterday by Gordon Atwater, curator of Hayden Planetarium, as responsible for the week's series of magnetic disturbances in radio. The largest spot ever seen in the history of the Planetarium occurred on March 10, 1947. The largest group of sunspots occurred about one month later, further influencing terrestrial magnetic conditions hence radio. Mr. Atwater said that the current spot, now disappearing over the sun's right-hand limb or edge, would appear again in some ten to twelve days over the sun's left-hand rim, and might cause as much trouble as this visitation. The rotational period of a sunspot is about 27 days. [New York Times, January 28, 1949 p. 42]

February 21, 1950 - Sun storm disrupts radio cable service – Transatlantic communications were disrupted off and on yesterday by electrical disturbances in the atmosphere. Western Union's cable division reported intermittent short stoppages that were worst in mid-afternoon but not as bad in late afternoon. Mackay Radio reported a lot of trouble with wireless communications from Europe but also reported that by late afternoon conditions were somewhat improved. Associated Press service from London was interrupted several times during the day. In Washington, the Bureau of Standards said a mammoth magnetic storm in the upper atmosphere of the earth is causing violent disruption of short wave radio communication. The disturbance which began about noon is likely to continue for thirty-six hours, the bureau reported. "The storm is the result of a great outpouring of matter from a large sunspot region on the face of the sun" the bureau explained, adding that it had predicted such a disturbance three weeks ago. Little effect is likely on standard FM and television broadcasts. [New York Times, February 21, 1950 p. 5]

The 1950 Korean War Storm

August 20, 1950 – While the Korean War was not yet eight weeks old a solar flare disrupted vital communications for half a day, as well as Associated Press news dispatches and other commercial traffic between the United States, Europe and South America. Beginning at Noon on Saturday, news dispatches from Korea and Tokyo were disrupted, and one can assume that military communications were in a similar state of affairs. In fact, the military not only confiscated Associated Press teletype and telephone equipment, but ordered the delivery of more commercial Press Wireless news equipment 'to provide some auxiliary communications for other points'. [New York Times, August 20, 1950, p. 5]

July 2, 1951 - Sky show seen over wide area but in only parts of city – No effects to

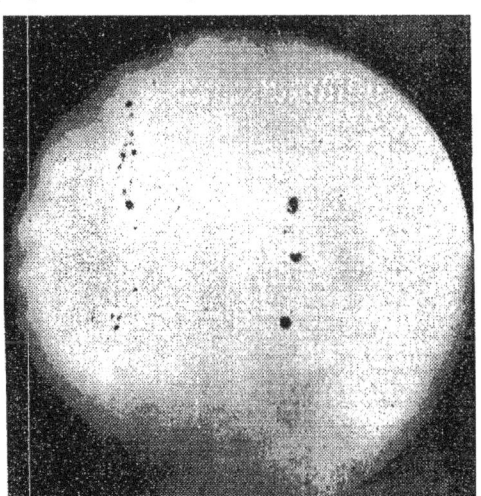

communications of the major broadcasting companies, Western Union or the Associated Press were reported by the electrical phenomenon. [New York Times, July 2, 1951, p. 10]. Northern lights visible over middle west area – The Associated Press reported that the display was visible over the northeastern United States from Wisconsin to New England. [Chicago Daily Tribune, July 2, 1951, p. C5].

Photo of sun by physicist E.P.Martz at Holloman Air Development Center, Alamogordo, New Mexico – (New York Times, Feb 26, 1956)

February 14, 1956 - Sunspots affecting radio links – A fairly large sunspot is seriously affecting world-wide radio communications and blackouts on some circuits are expected, British radio experts said here tonight. [New York Times, February 14, 1956 p. 59]

The 1956 Acheron Submarine Storm

February 24, 1956 : - The February 24, 1956 storm produced the most intense cosmic ray blast ever recorded, but was a nonevent so far as any apparent worldwide disruptions were concerned. [New York Times, February 24, 1956 p. 51] The February 24, 1956 storm, which produced the most intense cosmic ray blast ever recorded, seemed to be a non-event so far as any apparent world-wide disruptions were concerned. The New York Times ran a very short 100-word article on a 'rare aurora' seen in Fairbanks Alaska in the shape of a pure red arc. The New York Times ran a very short 100-word article on a 'rare aurora' seen in Fairbanks Alaska in the shape of a pure red arc [New York Times, February 26, 1956, p.44]. Later, they announced that 'Suns raging storms photographed [New York Times, February 25, 1956 p. 21] in a lengthier article.

BRITISH SUB (65 on board)
FEARED SUNK

LONDON, Friday: A full-scale air and sea search began in the North Sea today for the submarine Acheron believed sunk with 65 men aboard.

A sister ship, Affray, was lost in the English Channel in 1951, with 75 aboard.

Acheron, under-going trials in 145 fathoms off the east coast of Greenland, failed to make a signal check at 10 a.m. today, and the Admiralty flashed the emergency code "Submiss" to all warships.

This is the class of submarine reported missing.

Headline from The Argus newspaper, Melbourne, Australia for February 25, 1956.

But, the next day it was held responsible for the cause of a full-scale Naval alarm for a British submarine which was thought to have disappeared. The Acheron had been expected to report her position at 5:05 EST while on Arctic patrol. When it failed to do so, emergency rescue preparations were begun and both ships and rescue planes began searching the waters between Iceland and Greenland. The 'missing' submarine turned up four hours later when its transmissions were again picked up. [New York Times, February 25, 1956 p. 21]

Amidst a flurry of articles about US space launches, science education, and an advertisement for 'soldering guns' from Jordan March, on February 11 the front page of the Boston Globe ran a story "Sky Brilliance Among New England's Finest Shows in 30 Years". Channel 7 TV viewers were treated to broadcasts from Channel 7 in Manchester Vermont while Channel 4 viewers watched a program on the same channel televised from

Providence Rhode Island. A bizarre voice over to a love scene being played out on a local TV channel produced the dialog' *Smith gave him a left to the jaw and a short right to the stomach....But darling, we love each other so very much...a left hook to the jaw flattened Smith...Kiss me again my sweet*" Amazingly a local weatherman was quoted as saying that aurora were caused by sunlight reflecting off of ice crystals high up in the atmosphere, not realizing that he was resurrecting Captain John Rosse's discredited idea now dead for some 120 years.

March 16, 1956 - Radio disrupted by huge sunspot – A huge gaseous flare-up on the sun blacked out international wireless communications over half the earth for almost an hour yesterday. Overseas short-wave circuits and ship-to-shore communications were disrupted. Radio and television broadcasting services were not affected. [New York Times, March 16, 1956 p. 21]

January 25, 1957 - Uranium and aurora blamed in plane crash. – Air force experts today blamed uranium deposits in the Pyrenees and the rarely seen aurora borealis for the crash of a French air force DC3 on a mile-high peak, with the probable loss of the nine men aboard. They said the uranium threw the plane's compass and other navigational instruments out of kilter, and blamed disturbing magnetic storms on the northern lights. [Chicago Daily Tribune, January 25, 1957, p. 7].

March 4, 1957 - New awesome lights hang in north skies – The aurora borealis flared all over Sweden last night and many railroad signals showed green when they were set for red. Officials attributed the phenomenon to cosmic radiation. [Chicago Daily Tribune, March 4, 1957, p. 11].

April 17, 1957 - World radio signals fade – A spectacular fadeout of radio signals all over the world occurred today, starting at about 5 A.M. Eastern Standard Time and lasting from a few minutes to several hours in some areas. This type of fade-out is named for its discoverer – an American called Dellinger- and is not uncommon during periods of high sunspot activity such as are being experienced at present. [New York Times, April 18, 1956 p. 25]

June 28, 1957 and July 1, 1957 - The International Geophysical Year began appropriately enough with a cosmic curtain raiser in which the sun itself was the star performer...As was expected the first flare caused disruptions in radio communications and resulted in a severe radio blackout in Britain for more than thirty-six hours...The solar flares are believed to cause exceptional emissions of ultraviolet and X-rays in the ionosphere, the atmospheric layer some fifty to 400 miles above the earth. These rays lead to the formation of an electric blanket on the underside of the ionosphere, which allows short radio waves to escape into outer space instead of being reflected back to earth. [New York Times, July 7, 1957, p. 119]

September 5, 1957 - Aurora borealis stages sky show in Chicago area [Chicago Daily Tribune, September 5, 1957, p. 1].

September 13, 1957 - Rare northern lights display in southland [Los Angeles Times, September 13, 1957, p. 1].

September 23, 1957 - Aurora lights northern sky in city region [Chicago Daily Tribune, September 23, 1957, p. 2].

November 6, 1957 - Radio and TV, Sunspots in high gear. Sound of BBC video fills US homes - [New York Times, November 6, 1957 p. 71]

6

Storms in the Space Age

Although the USSR managed to surprise the United States by orbiting Sputnik 1, our entry into the Space Age came in 1958 with the launch of the Explorer 1 satellite. The main objective of the satellite was simply to staunch the perception that we had fallen behind the USSR in a critical technological area. So the satellite, no bigger than a large beach ball, was put on the engineering fast track and equipped with a simple experiment devised by James Van Allen at the University of Iowa. Even before the first satellite entered the space environment, scientists had long suspected that there would be some interesting things for instruments to measure when they got there. What they couldn't imagine was that billions of dollars of satellite real estate would eventually fall victim to these same cosmic bullets.

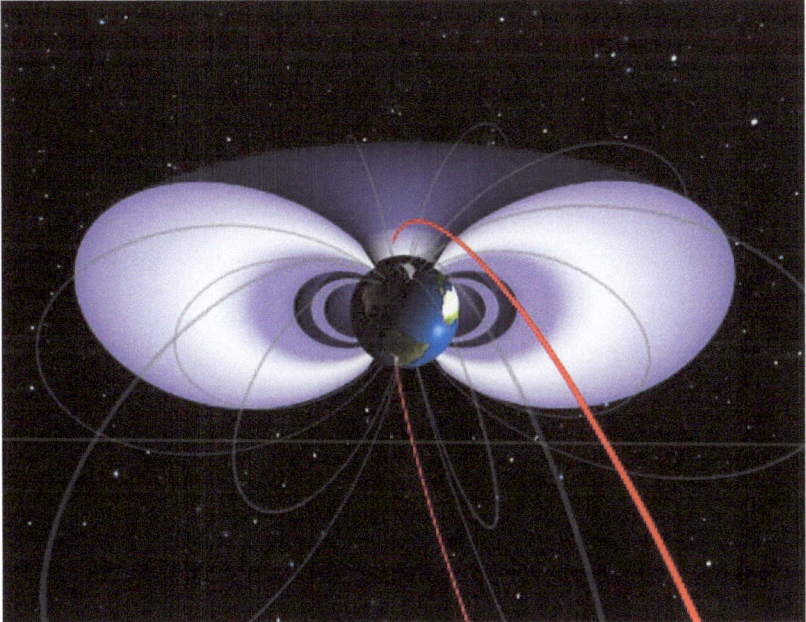

Purple and white areas represent the radiation belts. The red ring represents the orbit of
NASA - IMAGE satellite (Credit: NASA/Tom Bridgman)

More than ten years earlier, physicists working with photographic films on mountaintops had detected a rainstorm of 'cosmic rays' streaming into the atmosphere, but their origins were unknown. Van Allen wanted to measure how intense this rain was before it was muffled by Earth's blanket of atmosphere, and perhaps even sniff out a clue about where they were coming from in the first place. His experiment was nothing more than a Geiger Counter tucked inside the satellite, but no sooner was the satellite in space but the instrument began to register the clicks of incoming energetic particles. Space was indeed 'radioactive'. Since then, the impact that these particles have had on delicate satellite electronics has been well documented by civilian and military scientists. Even without any solar storms in progress the 'Van Allen radiation belts' are a magnetic bottle for high-energy particles that can zap delicate satellite electronics and provide a daily hazard to satellite operations. By 2014 we have over 1000 operational satellites orbiting Earth, providing everything from sensitive military intelligence and credit card verification, to the Home Shopping Channel. As with all previous technologies, our vulnerability to space weather and solar storm events in satellite technology depended on how many satellites were being used, and how many people they impacted. Also, many companies preferred not to publicize why their satellites 'mysteriously' stopped working or suffered temporary outages!

February 11, 1958 - Radio blackout cuts US off from the rest of the world. Aurora visible in Los Angeles, Tulsa, Boston, Seattle, Canada and Newfoundland. Voltages in electrical telegraph circuits exceeded 320 volts in Newfoundland. Intense red glow gave way to curtains and shimmering draperies....The aurora was accompanied by an electric storm that ended all radio communications between the United States and other countries and that disrupted telephone, teletype and electric circuits. [New York Times, February 11, 1958, p. 62]. Although not seen over New York, it was so intense over Europe that people wondered about fires and warfare. The storm virtually isolated the United States from radio contact with the rest of the world. [New York Times, February 12, 1958, p. 16]. Aurora puts on display in northern skies [Chicago Daily Tribune, February 11, 1958, p. 4]. Skies brilliant in northern lights display [Los Angeles Times, February 11, 1958, p.1]. Aurora borealis again seen here [The Washington Post, February 11, 1958, p. A1].

March 29, 1959 - Aurora seen on Long island, 2 hour display observed, radio disruption goes on – The sunspot flares occurred Tuesday and have severely disrupted radio communications across North Atlantic. RCA Communications said conditions had improved last night but were still far from ideal. The interference is expected to clear by tomorrow. [New York Times, March 29, 1959, p. 33]

July 16, 1959 - Radio upset by magnetic disturbance – A magnetic storm described by the air force Cambridge research center as extremely violent Wednesday caused aurora borealis and electrical ground currents which disturbed radio, teletype and similar communications systems in a wide area of the northeast. Coinciding with the magnetic storm was unusual sunspot activity on the sun. Telephone company reports showed the disturbance extending southward at least to Baltimore. The magnetic storm virtually blacked out radio communications with Europe throughout the day. Cable communications also were disrupted as communications normally sent by radio were switched to cables. Delays of up to four hours in cable transmission were reported. Associated Press reception of news from Europe was affected. The New England Telephone and Telegraph company reported it measured peak ground currents, induced by the storm, that ran as high as 150 volts.[Chicago Daily Tribune, July 16, 1959, p. C9]

November 29, 1959 - Aurora borealis seen in Houston [Los Angeles Times, November 29, 1959, p. A4].

April 1, 1960 - Aurora borealis viewed here [The Washington Post, April 1, 1960, p. A1].

October 7, 1960 - Sky in area is colored by northern lights [New York Times, October 7, 1960, p. 68].

November 13, 1960 - Type 3 solar flare gives North America a rare auroral display. – Solar radiation bombarding the earth's atmosphere at speeds of 3,000 feet/second caused magnetic storms around the world that washed out radio communications. An American Telephone and Telegraph spokesperson here explained that radio waves are beamed into the ionosphere during transmission. After the solar bombardment the disturbed ionosphere absorbed the waves instead of reflecting them. [New York Times, November 14, 1960 p. 14]. Display of northern lights here creates glow [New York Times, November 13, 1960, p. 3]. Solar explosion causes show of northern lights [Chicago Daily Tribune, November 14, 1960, p.1]. Blasts on sun roil earth's radio waves [Chicago Daily Tribune,

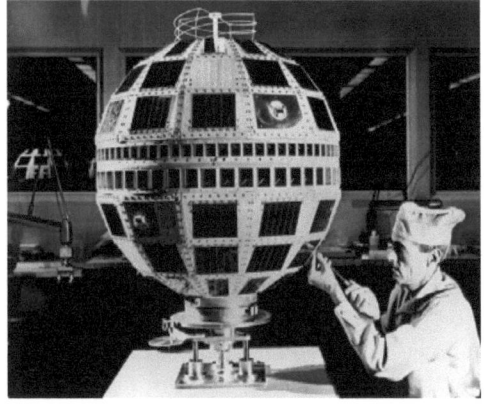

November 16, 1960, p. 16]. Aurora borealis proves thriller [The Washington Post, November 13, 1960, p. A1]. Aurora borealis lights up D.C. Area; Resultant calls light switchboards [The Washington Post, November 14, 1960, p. A3].

Telstar 1 being assembled (Credit: Bell Labs)

November 1962 - The first satellite to fall victim to space weather effects was, in fact, the one of the first commercial satellites ever launched into orbit in July 1962: Telstar 1. In November of that year, it suddenly ceased to operate. From the data returned by the satellite, Bell Telephone Laboratory engineers on the ground tested a working twin to Telstar by subjecting it to artificial radiation sources, and were able to get it to fail in the same way. The problem was traced to a single transistor in the satellites command decoder. Excess charge had accumulated at one of the gates of the transistor, and the remedy was to simply turn of the satellite for a few seconds so the charge could dissipate. This, in fact, did work, and the satellite was brought back into operation in January, 1963. The source of this information was not some obscure technical report, or an anecdote casually dropped in a conversation. This example of energetic particles in space causing a satellite outage was so uncontroversial at that time, it appeared under the heading 'Telstar' in the 1963 edition of the World Book Encyclopedia's 1963 Yearbook.

May 27, 1967 - City gets rare look at northern lights [The Washington Post, May 27, 1967, p. B3].

May 24, 1969 - Aurora borealis seen from N.Y to Louisiana [Los Angeles Times, May 24, 1969, p. A5].

The 1972 Apollo 17 'Near Miss' Storm

Hydrogen-alpha image of the August 1972 solar flare. The white area is the site of the flaring activity. (Credit: NOAA)

August 2, 1972 - - Solar astronomers reported that Active Region 331 had produced three powerful flares during a span of 15 hours. The intensity of these flares, classified as 'X2' were near the limits of the scale used to classify solar flare X-ray power. The next day, the Pioneer 9 spacecraft detected a shock wave from the first of these flares at 11:24 UT accompanied by a sudden change in the solar wind speed from 350 to 585 km/sec. Space weather forecasters at the Space Environment Services Center in Boulder Colorado issued an alert that predicted a major storm would arrive at the earth between August 4. They were not disappointed. Armed with vastly improved technology and scientific ideas, they were able to realize William Ellis's 1882 dream of predicting a solar storm. At 4:00 UT, aurora were seen simultaneously from Illinois to Colorado and the events of this storm were widely reported in major international newspapers. At 22:30 UT AT&T reported a voltage surge of 60 volts on their coaxial telephone cable between Chicago and Nebraska. Another 30 minute shutdown of phone service on Bell's cable link between Plano, Illinois and Cascade, Iowa was also attributed to the storm. Both the Canadian Overseas Telecommunications Corporation and Canadian National Telecommunications reported that the current surges in their lines had damaged components in their system ranging from noise filters to 'carbon blocks' Taxi drivers received orders from distant cities and were forced to turn down lucrative transcontinental fares! Paul Linger of the Denver Zoo said that the disruption of the Earth's magnetic field by the storms would disorient pigeons who depend upon the field for their sense of direction.

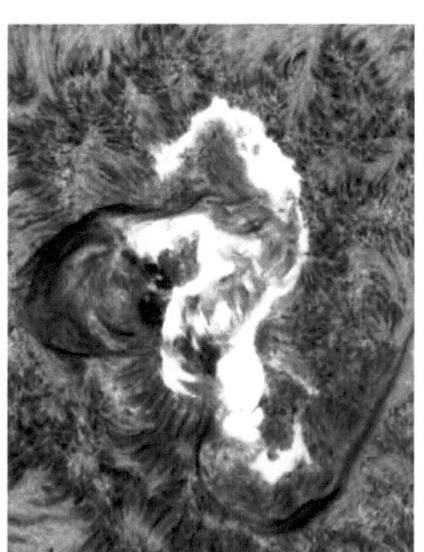

August 1972 'seahorse' flare captured in hydrogen-alpha light. (Credit: NOAA)

On August 2, 1972, solar astronomers reported that Active Region 331 had produced three powerful flares during a span of 15 hours. The intensity of these flares, classified as 'X2' were near the limits of the scale used to classify solar flare X-ray power. The next day, the Pioneer 9 spacecraft detected a shock wave from the first of these flares at 11:24 UT accompanied by a sudden change in the solar wind speed from 350 to 585 km/sec. Space weather forecasters at the Space Environment Services Center in Boulder Colorado issued an alert that predicted a major storm would arrive at the earth between August 4. They were not disappointed. Armed with vastly improved technology and scientific ideas, they were able to realize William Ellis's 1882 dream of predicting a solar storm.

At 4:00 UT, aurora were seen simultaneously from Illinois to Colorado and the events of this storm were widely reported in major international newspapers. At 22:30 UT AT&T reported a voltage surge of 60 volts on their coaxial telephone cable between Chicago and Nebraska. Another 30 minute shutdown of phone service on Bell's cable link between Plano, Illinois and Cascade, Iowa was also attributed to the storm. Both the Canadian Overseas Telecommunications Corporation and Canadian National Telecommunications reported that the current surges in their lines had damaged components in their system ranging from noise filters to 'carbon blocks'. Meanwhile, it was reported that some taxi drivers received orders from distant cities and were forced to turn down lucrative trans-continental fares! Paul Linger of the Denver Zoo said that the disruption of Earth's magnetic field by the storms would disorient pigeons who depend upon the field for their sense of direction.

Apollo 17 astronauts Harrison Schmidt and Eugene Cernan operating on the lunar surface in December 1972. Had they been there during the August 4 flare, they would have been exposed to high radiation levels.

At 1:20 AM Eastern Daylight Time on August 4, 1972, the Sun let-loose one of the most powerful blasts of radiation ever recorded during the Space Age. The streams of

X-rays and high-energy protons that flowed past the Earth within minutes, but not before triggering a major geomagnetic disturbance disrupting telephone service, and destroying a power transformer at the British Columbia Hydro and Power Authority. Although ground-based observers were kept on their toes by the unexpected power and communication outages, the event would have had a much more deadly outcome had it arrived four months later between December 7-19 while Apollo 17 astronauts were outside their spacecraft playing golf. Within a few hours, some estimates suggest that Harrison Schmidt and Eugene Cernan would have been hit by an incredible blast of radiation well over 1000 rem.

The astronauts would have suffered acute radiation sickness by the time they reached their Lunar Ascent Module, and probably even died some time later back on Earth. This is why James Michener, in his book 'Space' dwells on a similar event in his story of the fictional Apollo 18 mission. Some experts down play what a flare like the 'Apollo 17' flare might have actually done. Gordon Woodcock, for example, writes in his book 'Space Stations and Platforms' that *"Had an Apollo crew been on the lunar surface during the 1972 flare, they would very likely have received enough radiation to become ill. Radiation sickness effects at an exposure level of a few hundreds of rem take hours or days to become debilitating. James Michener's description in Space was not accurate"* Others beg to differ. According to Alan Tribble's, The Space Environment: Implications for spacecraft design, *"During August 1972 and again in October 1989, there were two extremely large solar proton [flares]. If an astronaut had been on the Moon, shielded by just a space suit, the radiation dose would probably have been lethal!"*

The orbiting Command Module would not have altered the outcome significantly according to shielding calculations by physicist Lawrence Townsend and his collaborators at NASA's Langley Research Center. Their 'worst case' analysis shows how the August 1972 flare radiation would have punched through bulkheads similar to those in the Apollo mission, and given the astronauts dosages as high as 250 rems, "Such an acute exposure would likely have incapacitated the crew because of radiation sickness and could possibly be lethal" Even this dosage is nothing to be sanguine about. Most radiation dosage tables say that 20% of the people exposed to even this level are sure to die within a month or two.

The Great Aurora of August 2, 1972, triggered surges of 60 volts on AT&T's coaxial telephone cable between Chicago and Nebraska. Meanwhile, the Bureau of Reclamation power station in Watertown, South Dakota experienced 25,000-volt swings in its power lines. Similar disruptions were reported by Wisconsin Power and Light, Madison Gas and Electric, and Wisconsin Public Service Corporation. The calamity from this one storm didn't end in Wisconsin. In Newfoundland, induced ground currents activated protective relays at the Bowater Power Company. A 230,000-volt transformer at the British Columbia Hydro and Power Authority actually exploded. The Manitoba Hydro Company recorded 120-megawatt power drops in a matter of a few minutes in the power it was supplying to Minnesota.

Despite the stealthy mayhem visited on us by its Great Aurora, 1972 was actually a very good year for electrical power in North America. We had far more available power than we used even during peak load conditions in the summer. Air conditioners were still pretty rare even in the urban world. With each passing year, however, we have found more uses for electricity than the pace with which we have created new supplies for it. The advent of the personal computer alone has added more than 3,000 megawatts per year to domestic power consumption since the 1980's. Steadily, the buffer between load and demand has been whittled away. Solar and geomagnetic storms continue to happen, but now there is much less wiggle room for power utilities to find, and purchase, additional power to tide them over. We don't build new power plants with the fervor we used to during the Go-Go '60s. No one wants them in their community, and those ugly power towers 100 feet tall are anathema to our suburban esthetics. So now utilities have learned how to buy and sell dwindling reserves of available power across states and whole regions.

July 6, 1974 - Aurora borealis lights the sky [Chicago Tribune, July 6, 1974, p. F3].

April 29, 1978 – Atlantic Communications Broken by Largest Solar Flare in Years – Federal scientists reported the detection of a major solar flare this morning and said they expect it to disrupt communications worldwide this weekend. Coast Guard officials in New York said that the flare had already knocked out maritime communications in the Atlantic for an hour and 45 minutes this morning. The disruptive effects of the flare, which reached a maximum intensity at 8:08 AM, are expected to reach the earth in the form of a magnetic storm on Sunday…some observers of the flare had set its radio emissions at the largest since 1946, although the emissions of X-rays were not believed to be as great. "The flare was on the eastern side of the sun at a place that usually means minimal interference with earth communications, but the size was such that we believe that there will be disruptions" said Dr. Gary Heckman of the National Center for Atmospheric Research…The flare was first detected by sensors aboard the GOES-2 satellite of the National Oceanographic and Atmospheric Administration this morning… Lieutenant Commander Thomas Osborne at the Coast Guard Station on Governor's Island, New York said 'There was a very large communications drop out between 8:30 and 10:15 this morning." After it was over, ships had been ordered to operate their communications at higher frequencies that better enables them to overcome the interference. [New York Times, April 29, p. 8]

April 13, 1981 - Aurora borealis lights up sky [Los Angeles Times, April 13, 1981, p. 1].

November 27, 1982 – Residents of some states might be treated to a display of northern lights late Sunday because of a major solar flare detected on Thanksgiving Day, Government scientists reported today. The display is also expected to cause some disruptions of communications systems, said Frank Guy, a solar forecaster at the joint NOAA-Air Force Space Environment Services Center. Mr. Guy said that as a result of the flare, the lights or the aurora borealis might be visible in such states as Montana, North Dakota and New York. [New York Times, November 27, p. 6]

Satellites in space have to be very carefully designed to avoid damage from high-speed and very high-energy particles, which can damage sensitive circuitry and cause satellite anomalies and outages of service. (Credit: NASA / RHESSI)

7 Storms in the Satellite Era

NASA's orbiting Solar Maximum Mission spacecraft gave us our first ring-side seat to the active sun. In particular, it let us see the material that was being ejected from the solar surface by creating an artificial total solar eclipse so that the fainter details could be observed. The Solar Maximum Mission (SMM or SolarMax) was launched on February 14, 1980. It carried several scientific instruments which provided new insights into the nature of solar flares. The spacecraft was rescued and repaired by a 1984 Space Shuttle Challenger mission. SMM reentered the Earth's atmosphere and burned-up on December 2, 1989.

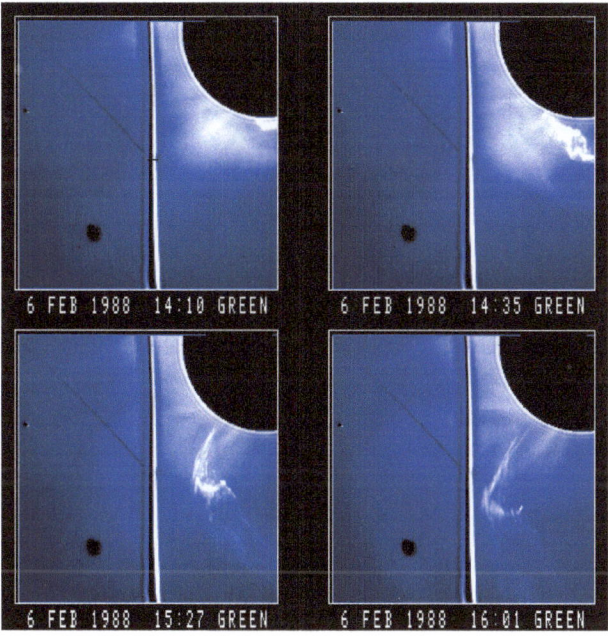

This panel of four images by the SMM satellite shows a 'puff' of plasma ejected on February 6, 1988. (Credit:NASA/SMM)

May 7, 1988 - Solar discharge sends storm over Earth – Storms like this can interfere with high-frequency radio transmissions, telephone communications and satellite communications. Mr. Gary Heckman chief of the Space Environment Services Center at NOAA said there had been few reports of communications interference so far, except for problems with one communications satellite. The SMM satellite recorded the ejection of mass Wednesday and that a small flare lasting more than three hours shot up from the sun before crashing back to the surface. The storm resulted in unusually bright aurora displays in the upper atmosphere and disturbed some long-distance communications according to NOAA. [New York Times, May 7, 1988, p. 36].

The 1989 Quebec Blackout Storm

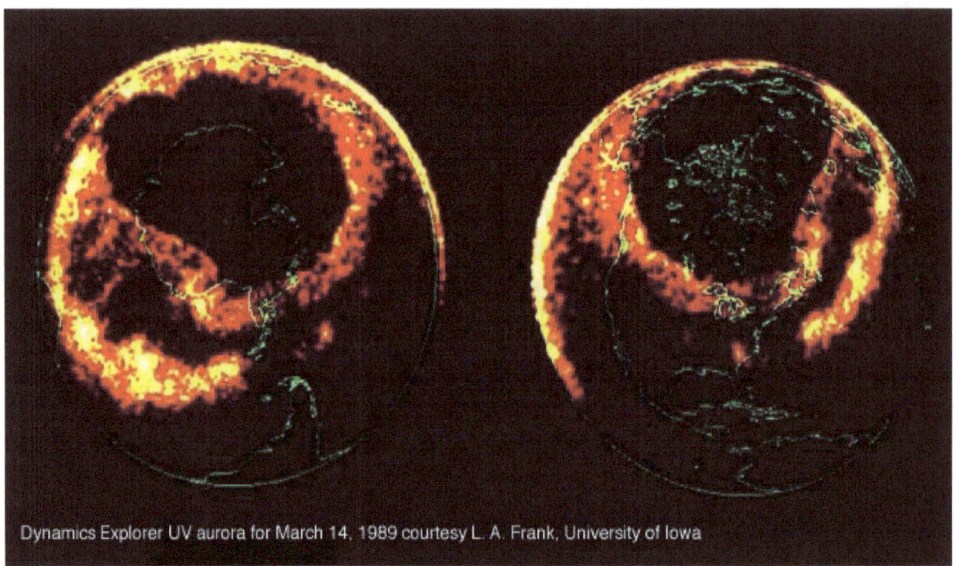

Dynamics Explorer UV aurora for March 14, 1989 courtesy L. A. Frank, University of Iowa

March 13, 1989 - Astronomers were busily tracking "Active Region 5395" on the Sun when suddenly it disgorged a massive cloud of superheated gas on March 10, 1989. Three days later, and seemingly unrelated to the solar paroxysm, people around the world saw a spectacular Northern Lights display. Most newspapers that reported this event considered the spectacular aurora to be the most newsworthy aspect of the storm. Seen as far south as Florida and Cuba, the vast majority of people in the Northern Hemisphere had never seen such a spectacle in recent memory. At 2:45 AM on March 13, electrical ground currents created by the magnetic storm found their way into the power grid of the Hydro-Quebec Power Authority. Giant capacitors tried to regulate these currents but failed within a few seconds as automatic protective systems took them off-line one by one. Suddenly, the entire 9,500 megawatt output from Hydro-Quebec's La Grande Hydroelectric Complex found itself without proper regulation. Power swings tripped the supply lines from the 2000 megawatt Churchill Falls generation complex, and 18 seconds later, the entire Quebec power grid collapsed. Six million people were affected as they woke to find no electricity to see them through a cold Quebec wintry night. People were trapped in darkened office buildings and elevators, stumbling around to find their way out. Traffic lights stopped working, Engineers from the major North American power companies were worried too. Some would later conclude that this could easily have been a $6 billion catastrophe affecting most US East Coast cities. All that prevented the cascade from affecting the United States were a few dozen capacitors on the Allegheny Network. [New York Times, March 13, 1989 p. B5]

During March 13-14 1989 flare and solar storm activity, a shock wave was produced that collided with the Earth's magnetosphere and compressed it inside the orbits of the geostationary satellites. LORAN navigation signals for pilots were interrupted so that ships and aircraft could not navigate properly for several hours.

Also during the March 1989 storm, a transformer at a nuclear plant in New Jersey was damaged beyond repair as its insulation gave way after years of cumulative GIC damage. Allegheny Power happened to be monitoring a transformer that they knew to be flaky. When the next geomagnetic storm hit in 1992. They saw the transformer reply in minutes, and send temperatures in part of its tank to more than 340 F (171 C). Other transformers have spiked fevers as high as 750 F (400 C). Insulation damage is a cumulative process over the course of many GICs, and it is easy to see how cumulative solar storm and geomagnetic effects were overlooked in the past.

"The General Motors car-assembly plant in Boisbraid lost production of $6.4 million worth of automobiles...The Montreal Stock Exchange, located in Place Victoria, was forced to operate on emergency power. Most trades had to be completed manually...Sidbec-Dosco, Inc., a Quebec-owned steel company, estimated yesterday's production loss at between $500,000 and $1.5 million, "All the steel that was already on the line in the hot rolling mills is scrap"... Cascades ,Inc., a pulp and paper company based in Kingsey Falls, said the power shutdown would cost his company between $200,000 and $300,000. The amount doesn't include salaries."

Core of 345 kV transformer in New Jersey damaged by the March 1989 geomagnetic storm. (Courtesy, Paul Kappenman).

"Radio Free Europe said yesterday that its engineers found no indication the Kremlin had resumed jamming it or its sister station, Radio Liberty, to block reports on demonstrations in the Soviet Union. Radio Free Europe spokesman Bob Redlich said that an effect similar to jamming could have been caused by recent increases in solar activity, which can hamper radio reception."

The 1989 Toronto Stock Market Storm

August 16, 1989 - A significant geomagnetic storm caused by a very large X20 solar flare affected microchips and lead to the halt of Toronto's stock market trading on August 16, 1989. The solar flare was stronger than the X15 flare recorded in March of the same year. Officials watched in disbelief as three disc drives failed in succession on what is supposed to be a 'fault-tolerant' computer system. The crash stopped stock market trading for three hours. 'I don't know what the gods were doing to us,' said John Kane, the vice president of the exchange. The Sun is currently unusually active. The present period of activity - marked by solar flares, discharges of charged particles and magnetic storms on Earth - began in the spring. Scientists in Canada believe that the Sun is responsible for malfunctions in many personal computers across the country and for a two-hour power cut in parts of the province of Quebec. [New Scientist, September 9, 1989]

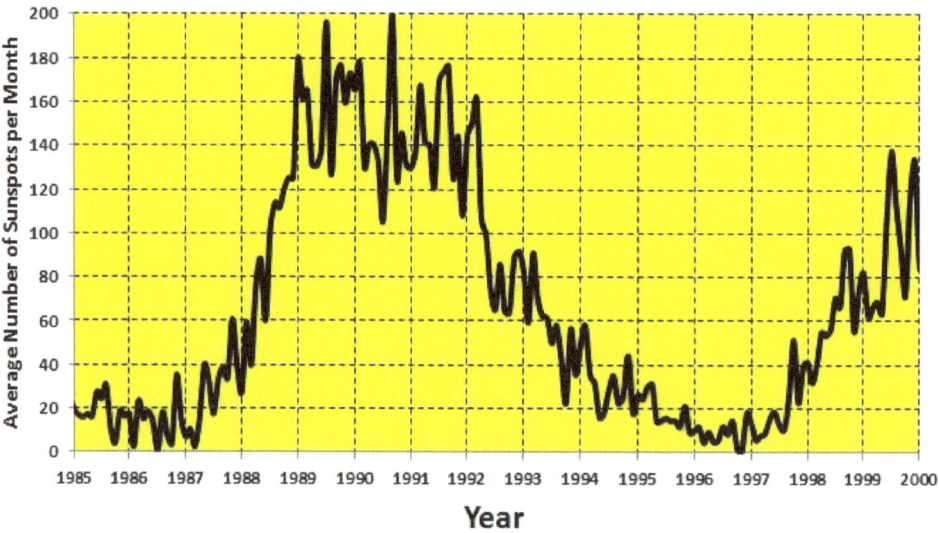

Sunspot cycle 22 began in 1986 and reached its peak between 1989-1992 before declining to a minimum again in 1997. The peak of the cycle, identified by the highest average number of sunspots seen in a given month, occurred in 1990.

November 22, 1989 - The Insat-1C communications satellite failed during the last of a series of major solar and geomagnetic storm events of this memorable year. These storms caused major increases in energetic protons at energies above 10 MeV, between October 19 through November 6, rivaling all of the events in solar cycle 22 taken together. On November 15th another powerful solar blast was detected on ground-level neutron monitors, and at the end of November another major solar flare, rich in high-energy protons, was recorded. According to space researcher Joe Allen at the NOAA National Geophysical Data Center, each of these events caused power panel degradation in a variety

of satellites. Some lost 5-7 years of usable lifetime. Others suffered a variety of glitches and operational anomalies.

March 25, 1991 - The Marecs-1 satellite suffered a complete failure, and had a long history of space environment problems. Its predecessor Marecs-A launched in December 1981 had already been disabled ten years earlier during a week of intense auroral activity in February 1982.

June 6, 1991 - Severe sun storm threatens utilities – A severe disturbance of the Earth's magnetic field caused by temporary changes in solar activity, began on Tuesday night, threatening electric utility equipment and communication systems, Government scientists said yesterday. The storm, which produced displays of the northern lights as far south as Pennsylvania on Tuesday, is expected to persist for several days and may intensify. Utility managers around the country were notified yesterday to remain on alert, because such storms can interrupt electrical transmissions and damage transformers at generating stations .Some officials have questioned whether explosions and a fire last month at the Maine Yankee nuclear plant in Wiscasset may have been caused by such a storm. The

current storm did not immediately threaten the crew of the space shuttle Columbia, launched early yesterday, but it might subject the astronauts to unusual levels of radiation later this week, one expert said. The dose would be too low to cause immediate health effects, he said, but could exceed Federal safety limits set for people who work with radiation. Airline passengers might also experience heightened dose, especially those flying near the poles. Some experts have said that on flights over the poles, the doses kin such storms could be of concern to pregnant women. The storm is also exposing some satellites to abnormally high levels of radiation, which may cause them to malfunction at least briefly. Interference with high-frequency radio transmission, including short-wave, citizen's band and AM radio waves was spotty according to forecaster Chris Balch at the Space Environment Center in Boulder. [New York Times, June 6, 1991, p. A16].

Satellites receive their operating power from large-area solar panels which have surfaces covered by solar cells. When the Sun ejects clouds of high-energy protons, these particles can literally scour the surfaces of these solar cells. Direct collisions between the high-speed protons, and the atoms of silicon in the cells, cause the silicon atoms to violently shift position. These shifting atoms produce crystal defects that increase the resistance of the solar cells to the currents of electricity they are producing. Solar cell efficiency steadily decreases, and so does the power produced by the solar panels. Engineers have learned to compensate for this erosion of power by making solar panels over-sized. This lets the satellite start out with extra capacity to cover for this steady degradation of electrical output. But this degradation doesn't happen smoothly over time. Like a sudden summertime hailstorm, the Sun produces unpredictable bursts of particles, which do considerable damage in only a few hours. During October 19-26, 1989 a series of powerful solar flares caused many satellites to experience about five years of solar panel degradation in just seven days. Satellites that were designed to last 10 years, were now expected to last only five before their panels could no longer provide full power. The

GEOS-7 weather satellite lost half of its mission lifetime in just this way, from a single solar flare in March 1989.

High-energy particles also do considerable internal damage to spacecraft. At the atomic scale, to an incoming proton, the walls of a satellite look more like a porous spaghetti colander than some solid wall of matter. When high-energy protons do manage to collide with atoms in the walls of the satellite, they produce sprays of secondary, energetic electrons that penetrate even deeper into the interior of the satellite, producing what engineers call 'Internal Dielectric Charging'. As the charging continues, eventually the electrical properties of some portion of the satellite breaks down and a discharge is produced. In a word, you end up with a miniature lightning bolt that causes a current to flow in some part of an electrical circuit it's not supposed to. As anyone who has inserted new boards into their PC can tell you, just one static discharge can destroy the circuitry on a board. Beyond actual physical damage, these particles can also change information stored in a computer's memory.

Microscopic current flows can flip a computer memory position from '1' to '0' or cause some components, or an entire spacecraft system, to switch-on when it is not supposed to. When this happens, it is called a 'Single Event Upset' or SEU, and like water they come in two flavors: hard and soft. A hard SEU actually does unrepairable physical damage to a junction or part of a microcircuit. A soft SEU merely changes a binary value stored in a device's memory, and this can be corrected by simply 're-booting' the device. Engineers on the ground cannot watch the circuitry of a satellite as it undergoes a discharge or SEU event, but they can monitor the functions of the satellite. When these change suddenly, and without any logical or human cause, they are called 'Satellite Anomalies'. They happen a lot more often than you will ever read about in the news media.

More readily available data on this problem can be had from government research and communication satellites because the information is, at least in principle, open to public scrutiny if you happen to know who to talk to, or can extract the information from thousands of technical reports.

TDRSS-1 satellite (Courtesy: NASA)

The first satellite in the NASA, Tracking and Data Relay Satellite System (TDRSS-1) was launched in April 1983, and from that time onwards, the satellite has been continuously affected by soft SEUs. The satellite anomalies affected the spacecraft's Attitude Control System, and like mosquitoes on a warm day, they remain a constant problem today. The SEUs have been traced to changes in the computer's RAM, and the most serious of these SEUs were considered mission-threatening. If left uncorrected, they could lead to the satellite tumbling out of control. Ground controllers have to constantly keep watch on the satellite's systems to make certain it keeps its antennas pointed in the right direction. This has become such an onerous task that one of the ground controllers, the late Don Vinson, once quipped, "If this [the repeated SEU's] keeps up, TDRS will have to be equipped with a joystick"

The problems with TDRSS-1 quickly forced NASA to redesign the next satellites in the series, TDRSS-3 and 4 (TDRSS-2 was lost in the Challenger accident), and the solution was fortunately very simple. In engineering-speak, "The Fairchild static, bi-polar 93L422 RAMS were swapped for a radiation-hardened RCA CMM5114 device based on a different semiconductor technology". Radiation-hardening is a complex process of redesigning microcircuits so that they are more resistant to the high-energy particles that pass through them. The result is that neither of the two new TDRSS satellites have recorded SEUs while during the same operation period, hundreds still cause TDRS-1 to rock and roll, keeping the satellites human handlers steadily employed for the foreseeable future.

The 1994 Anik Satellite Storm

January 20, 1994, was a moderately active day for the Sun. There were no obvious solar flares in progress, and no evidence for any larger-than-normal amounts of X-rays, but a series of coronal holes had just swept across the Sun between January 13-19th. According to the NOAA Space Environment Center, the only sign of unrest near the Earth was the high-speed solar wind from these coronal holes which had produced active-to-minor storm conditions in their wake. NASA's, SAMPEX satellite, was beginning to tell another, more ominous, story. There were now signs of energetic electrons near geosynchronous orbit, whose concentration were rising to a maximum. These particles came from the passage of a disturbance from the magnetotail region into the inner magnetic field regions around the Earth. Within minutes, the GOES-4 and GOES-5 weather satellites began to detect accumulating electrostatic charges on their outer surfaces. Unlike the discharge you feel after shuffling across a floor, there is no way that satellites can unload the excess charges they accumulate, and so they continue to build until the surfaces reach voltages of hundreds, or even thousands of volts.

Image Credit: L. J. Lanzerotti, Bell Laboratories, Lecent Technologies, Inc.

The Anik E1 and E2 satellites, owned by Telesat Canada, were a twin pair of GE Astro Space model 5000 satellites, weighing about 7000 pounds, and lunched into space in 1991. From their orbital slots on the equator 900 miles southwest of Mexico City, and 1,500 miles apart in space, they soon become the most powerful satellites in commercial use in all of North America. Virtually all of Canada's television broadcast traffic passed through the E2 transponders at one time or another. The E2 satellite provided the business community with a variety of voice, data, and image services. Despite some technical difficulties with the deployment of the Anik E2 antenna which dogged engineers for several months, the satellites soon became a reliable corner stone for North American commerce and entertainment.

Canadians eagerly awaited this satellite service because major cities were few and far between across Canada; a territory bigger than the United States. With hundreds of small towns, and only a few dozen major cities with television stations, the satellites quickly became the information lifeline for many parts of Canada. About 2,300 cable systems throughout Canada, and nearly 100,000 home satellite dish owners depended on these satellites to receive their programming. Far-flung newspapers relied on these satellites to beam their newspapers to distant printing presses to serve local communities. Most people thought the satellites would continue working until at least 2003, but on January 20, 1994 this optimism came to an end.

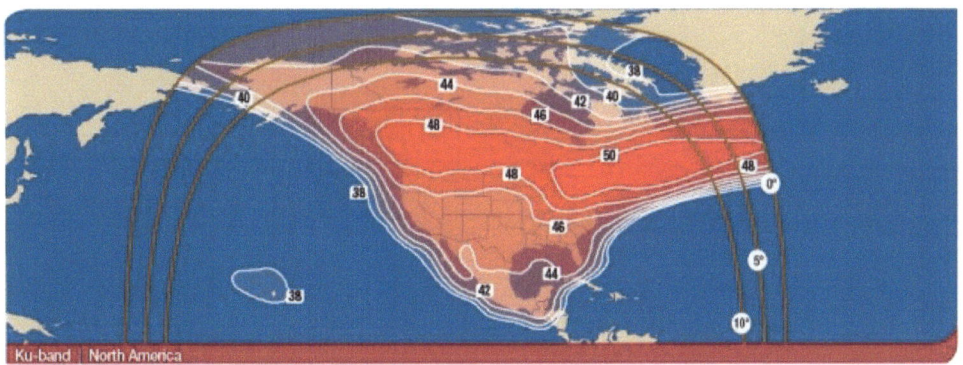

This map shows the area covered by the beam of the Anik-F2 satellite (Credit: TrackSats.com)

As the GOES satellites began to accumulate electric charges from the influx of energetic particles, the Intelsat-K satellite began to wobble on January 20, 1994, and experienced a short outage of service. About two hours later, the Anik satellites took their turn in dealing with these changing space conditions, and did not do as well. The satellites experienced almost identical failures having to do with their momentum wheel control systems. The first to go was Anik E1 at 12:40 PM which began to roll end-over-end uncontrollably. The Canadian Press was unable to deliver news to over 100 newspapers and 450 radio stations for the rest of the day, but was able to use the Internet as an emergency back-up. Telephone users in 40 northern Canadian communities were left without service. It took over seven hours for Telesat Canada's engineers to correct Anik E1's pointing problems using a back-up momentum wheel system.

About 70 minutes later at 9:10 PM, the Anik E2 satellite's momentum wheel system failed, but its backup system also failed, so the satellite continued to spin slowly, rendering it useless. This time, 3.6 million Canadians were affected as their major TV satellite went out of service. Popular programs such as MuchMusic, TSN and the Weather Channel were knocked off the air for three hours while engineers rerouted the services to Anik E1. For many months, Telesat Canada wrestled with the enormous problem of trying to re-establish control of Anik E2. They were not about to scrap a $300 million satellite without putting up a fight. After five months of hard work, they were at last able to re-gain control of Anik E2 4 on 21 June 1994. The bad news is that, instead of relying on the satellite's now useless pointing system, they would send commands up to the satellite to fire its thrusters every minute or so to keep it properly pointed. This ground intervention would have to continue until they ran out of thruster fuel, shortening the satellites lifespan by several years. The good news is that Telesat Canada became the first satellite company to actively stabilize a satellite without using any satellite attitude system. In the end, it would turn out to be something of a Pyrrhic victory because on March 26, 1996 at 3:45 PM, a crucial diode on the Anik E1 solar panel shorted out, causing a permanent loss of half the satellite's power. Investigators later concluded that this, too, was caused by an unlucky solar event.

The connection between the geomagnetic disturbance and the Anik satellite outages seemed to be entirely straight-forward to the satellite owners at the time, and

Telesat Canada publicly acknowledged the cause-and-effect relationship in press releases and news conferences following the outages. They also admitted that the Anik space weather disturbance which had ultimately cost their company nearly $5 million to fix, was consistent with past spacecraft-affecting events they had noticed and that very similar problems had also bedeviled the Anik-B satellite 15 years earlier. What also made this story interesting is that the Intelsat-K and the two Anik satellites are of the same satellite design. The crucial difference however, is that the Intelsat Corporation specifically modifies its satellites to survive electrostatic disturbances including solar storms and cosmic rays. This allowed the Intelsat-K satellite to recover quickly following the storms that disabled the unmodified Anik satellites. Clearly, it is possible, and desirable, to 'harden' satellite systems so that they are more resistant to solar storm damage. This lesson in spacecraft design is not a new one we have just learned, but a very old one that has been applied more or less conscientiously since the dawn of the Space Age itself when these problems were first uncovered.

The 1997 Telstar Satellite Storm

January 7, 1997 seemed to be an ordinary day on the Sun. Photographs taken at the Mauna Kea Solar Observatory showed nothing out of the ordinary. In fact, to the eye and other visible wavelength instruments, the images showed nothing at all. Not so much as a single sunspot. But X-ray photographs taken by the Yohkoh satellite from Earth orbit revealed some serious trouble brewing. High above the solar surface in the tenuous atmosphere of the corona, invisible lines of magnetic force, like taught rubber bands, were coming undone within a cloud of heated gas. Balanced like a pencil on its point, it neither rose nor fell as

The SOHO satellite observed the ejected plasma from the sun, called a coronal mass ejection or CME, as a halo event, which means it was directed at Earth on January 6.

magnetic forces levitated the billion-ton cloud high above the surface. Then, without much warning, powerful magnetic fields lost their anchoring and snapped into new shapes; their precarious balance between gravity and gas pressure, lost.

The massive cloud launched from the Sun, crossed the orbit of Mercury in less than a day. By Wednesday this cloud had passed Venus: An expanding cloud over 30 million miles deep, spanning the space within much of the inner solar system between the Earth and Sun. At a distance of one million miles from Earth, the leading edge of the invisible cloud finally made contact with NASA's WIND satellite at 8:00 PM EST on January 9. By 11:30 PM the particle and field monitors onboard NASA's earth-orbiting POLAR and GEOTAIL satellites told their own stories about the blast of energetic particles now sweeping through the solar system. Interplanetary voyagers would never have suspected the conflagration that had just swept over them. The cloud had a density hardly more than the best laboratory vacuums.

The tangle of fields and plasma slammed into the Earth's own magnetic domain like some enormous sledgehammer as a small part of the million mile-wide cloud brushed by Earth. Nearly a trillion cubic miles of space were now involved in a pitched battle between particles and fields, shaking Earth's magnetic field for over 24 hours. The storminess in space rode the tendrils of Earth's field all the way down to the ground in a barrage of activity. Major aurora blazed forth in Siberia, Alaska and across much of Canada during this long winter's night.

The initial blast from the cloud (astronomers call it a 'coronal mass ejection' or CME), compressed the magnetosphere and drove it inside the orbits of geosynchronous satellites, amplifying trapped particles to high energies. Dozens of satellites positioned at fixed longitudes along Earth's equator like beads on a necklace, alternately entered and exited the full-bore of the solar wind every 24 hours as they passed outside of Earth's magnetic shield. Plasma analyzers developed by Los Alamos Laboratories, and piggy-backed on several geosynchronous satellites, recorded voltages as high as 1000 volts, as static electric charges danced on their outer surfaces. It was turning out to be not a very pleasant environment for these high-tech islands of silicon and aluminum.

The storm conditions continued to rage throughout all of January 10th. Just as the conditions began to subside, on January 11th, Earth was hit by a huge pressure pulse as the trailing edge of the cloud finally passed by. The arrival and departure of this cloud would have not been of more than scientific interest, had it not also incapacitated a $200 million communications satellite in its wake at 06:15 EST: Telstar 401.

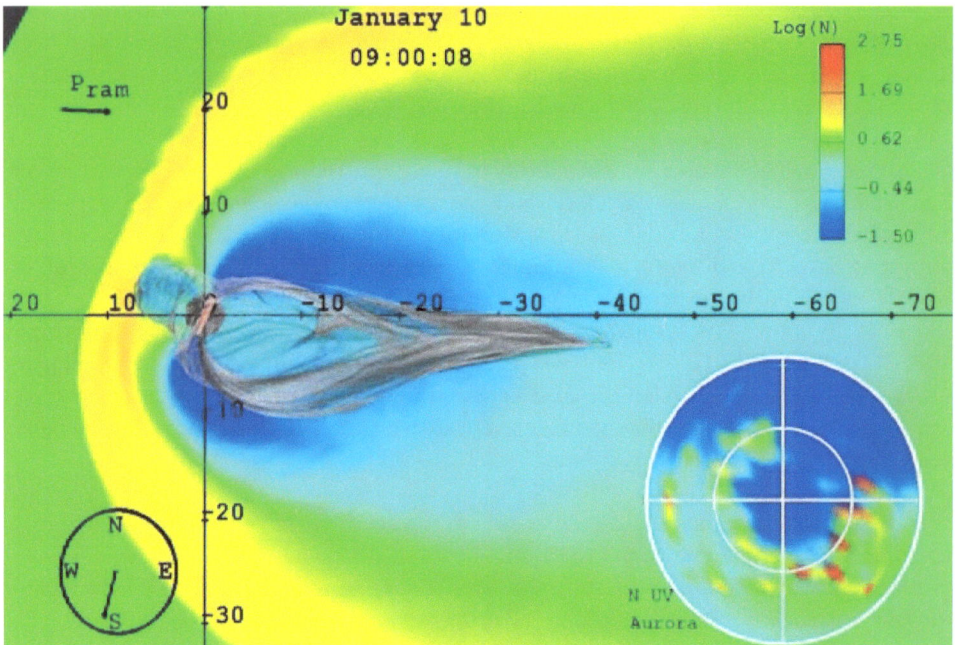

The Earth's magnetosphere as depicted by a computer model, showing a geomagnetic storm in January 1997. (Courtesy: Space Plasma Physics group at the University of Maryland).

AT&T tried to restore satellite operations for several more days, but on January 17th they finally admitted defeat and decommissioned the satellite. All TV programs such as 'Oprah Winfrey', 'Bay Watch', 'The Simpsons', and feeds for ABC News, had to be switched to a spare satellite, Telstar 402R. The Orlando Sentinel on January 12 was the first newspaper to mention the outage in a short, 74-line, note on page 22. Three days later, the Los Angeles Times described how this outage had affected a $712 million sale of AT&Ts Skynet telecommunications resources to Loral Space and Communications Ltd. No papers actually mentioned a connection to solar storms until several weeks later on January 23, when the focus of the news reports in the major newspapers was the thrilling scientific studies of this 'magnetic cloud' . The New York Times closed their short article on the cloud by mentioning that,"...*Scientists said they do not know if this month's event caused the failure early on January 11, of AT&T's Telstar 401 communications satellite, but it occurred during magnetic storms above the earth.*"

The 1997 Tempo-2 Satellite Storm

April 10, 1997 - Earth experienced a near-miss by a coronal mass ejection; an event that was followed in great detail by the news media thanks to the dramatic visual imagery from the NASA Solar and Heliospheric Observatory (SOHO). A month after the near-miss, Space News, a much-read weekly newspaper of the space community, carried a short article about a major new-generation satellite, which had encountered space weather difficulties at about the time this solar event reached Earth's orbit. The Tempo-2 satellite, equipped with the latest in high-power, gallium-arsenide solar arrays lost 15% of its operating power on

April 11, and this was directly credited to the solar storm by Loral's Space Systems Division who manufactured the satellite. In a May 1 statement from company spokesperson David Benton announced that, *"We have evidence from sensors on the satellites that there was a space event in the vicinity of the Tempo satellite at the time of the disturbance."*

The Space News article also mentioned that a spokesman for the satellite owner, TCI Satellite Entertainment, took a far more cautious position for why the satellite lost some of its operating power. Unlike Loral who credited the space weather event with the problem, TCI announced that they were *"unclear whether the storm had actually exceeded the levels the satellite was built to withstand, or if the satellite simply had a flaw"*. Compared to the murky causes surrounding the Telstar 401 outage, this level of candor by Loral was refreshingly to the point, even though the satellite owner preferred a more guarded opinion. It would, of course, be the satellite owner that would seek insurance payments, not the satellite manufacturer. In a replay of the Telstar 401 settlement, TCI filed a claim for $20 million.

Far from being just another satellite stamped from a tried-and-true design, Tempo-2 was supposed to be the vanguard of a whole new fleet of high-capacity communications satellites. Communication satellites had evolved from humble 10-watt 'small reptiles' to leviathan multi-kilowatt 'dinosaurs' driven by the relentless evolutionary pressure of consumer demand. To generate the 10s of kilowatts of power needed to operate dozens of transponders and other high-end equipment, engineers had been forced to create lighter-weight, and higher-power solar cells. The current darling for this new technology is based on semiconductor compounds of gallium and germanium, rather than the common silicon cell materials. The new cells would be wired to produce 60 volts per module to keep the weight and size of the solar panels within the limits set by the cost of the satellite. One of these panels, incidentally, could comfortably supply the needs of a medium-sized house. But the Tempo-2 failure uncovered a potentially fatal problem with these new panels. They were susceptible to energetic particle impacts, which caused miniature lightning bolts to flare-up and short circuit sections of the panels. Engineers would certainly have to go back to the drawing board to fix this problem, because these satellites were to be the wave of the future.

The 1997 Adeos Satellite Storm

September 20, 1997 - Although there is no public data on anomalies experienced by commercial or military satellites at this time, deteriorating space weather conditions by the end of September were openly cited as the cause of a Japanese satellite glitch. On September 20, 1997, the $474 million Adeos research satellite, launched by Japan a year earlier, began to malfunction. According to a report in Space News, *"...Cosmic rays were found to have damaged the main on-board computer, which caused it to shut down all non-essential systems, including the sensors, forcing scientists to reprogram its software"*

There were no obvious geomagnetic storms in progress near the Earth on this particular day according to the Canadian CANOPUS magnetic observatory network, so whatever space event had taken its toll on Adeos, it had managed to skirt the sparse network of instruments available to record it. When you have one trillion cubic miles to cover, there are plenty of opportunities for a handful of instruments to be in the wrong place to see anything. In many ways, it's like trying to keep track of the weather in Tokyo from monitoring stations in Stockholm, Manila and Rio de Janero.

The 1997 Insat Satellite Storm

September 27, 1997 - Close on the heels of the Adeos satellite problem, the Sun decided to get back into the act of terrorizing Earth. The SOHO satellite witnessed two major CME events on September 24 and September 27. The events were echoed in the data returned by the ACE satellite on September 30th as the CME plasma rushed by the satellite at nearly one million miles per hour. CNN News and the Reuters News Service reported that India had lost an advanced communications satellite, INSAT-2D on October 2, 1997 because of a power failure. The satellite, launched June 4, 1997, carried 24 transponders for relaying Indian telephone and television traffic. The satellite's problems seem to have started on, or before, October 1 when it lost Earth-lock briefly, shortly after the September 27th CME had passed Earth on September 30th. One of its predecessors, INDSAT-1C launched on July 21, 1988, fell silent under similar circumstances in 1989. According to a summary for INSAT-1C in Janes Space Directory *"..a power system failure from a solar array isolation diode short"* caused the satellite to lose half of its capacity. On November 22, 1989 the satellite lost its Earth lock and was abandoned at a cost of $70 million.

For INSAT-2D, ACE magnetometer data showed a sharp rise in solar wind strength on October 1 at 0000 UT followed by a persistent plateau of magnetic field intensity lasting a full day before subsiding again. The Earth-orbiting GEOTAIL satellite also noted a sharp change in the local energetic particle conditions as well as geomagnetic field strength. All of these are consistent with the arrival of the September 27 CME around October 1 at the time the INSAT-2D began having its problems.

April 29, 1998 - A series of low-level, X-class flares and multiple CMEs were expelled

1998/04/29 17:27

between April 29 and May 2. The outfall finally started to reach Earth around May 2-3 causing a severe geomagnetic storm. Only three times before, during the January 1997, April 1997 and November 1997 encounters, had scientists gone so far as to provide formal press briefings for impending calamities. Like earthquake forecasting, it is better to miss a few quietly, than announce false positives. But the conditions, this time, seemed to warrant some kind of official comment just in case the storms might grow to become something more provocative than the topic of a scientific

CME on April 29 imaged by SOHO

research paper. So far, the scientists seemed to be batting 100. The last three press releases had confirmed that scientists could anticipate the likely geospace impact of some solar storms. This new one would extend this winning streak to four.

On May 4 at 00:18 EDT a strong geomagnetically-induced currents (GICs) affected the northeast United States as capacitor banks were tripped. This caused transformer saturation affecting a major electrical substation in New Brunswick, and caused voltage regulation problems throughout Maine. Minutes later in New York state, voltages also started to drop in the eastern part of the state. The Nova Scotia electrical company measured 70 ampere GICs in one of its transformers. A day later, routine testing of a transformer in the Hudson Valley indicated insulation damage and a temperature spike of several hundred degrees Celsius. According to John Kappenman at MetaTech Corporation, the Hudson Valley electrical utility recorded this problem as due to 'undetermined causes'. Satellite owners had also experienced their own spate of problems, many of them fatal. The first satellite to succumb to these conditions, and publicly acknowledged, was the German research satellite Equator-S. On May 1 the satellite owned by the Max Planck Institute lost its backup processor. According to an announcement by the Institute at the official web site for this satellite, *"If a latch-up caused by penetrating particle radiation was the cause, there is hope that it may heal itself upon the next eclipse because of the complete temporary switch-off of the electrical system"*

The 1998 Pager Satellite Storm

May 19, 1998 – *Satellite Outage Felt By Millions - Not Just Pagers Were Affected As Radio And TV Networks And Various Other Businesses Using Galaxy IV Scrambled To Stay Connected.* – Computer failure aboard the $250 million, Galaxy IV telecommunications satellite sent hospitals, radio and television networks, businesses big and small, and millions of ordinary people scrambling to find alternatives to pagers, automated tellers and other satellite-dependent equipment. The satellite outage, which began late Tuesday and may not be fully repaired until next week, touched every corner of the country, knocking out the paging systems used by up to the 90 percent of Americans who carry pagers, according to Scott Baradell of PageNet Inc., of Dallas, which has 10 million pager customers. National Public Radio and other broadcasters had to find alternative means to feed programming to local stations, and animated weather maps were missing from many televised weather reports. Probably the most serious potential impact of the outage is at hospitals, which rely on pagers to contact physicians in emergencies. "No pagers, no doctors," lamented Chris Ortmann, a unit secretary in the bustling emergency room of Alexian Brothers Medical Center in Elk Grove Village. [Chicago Tribune, May 21, 1998]

When the Galaxy IV satellite ceased operating on May 19, 1997, millions of pager owners woke up the next day to discover that their high-tech devices had turned into useless pieces of plastic. When they got into their cars and tried to pump gas at the local gas station, the pumps rejected their credit cards because they were unable to use the satellite to transmit and receive verification codes. 100,000 privately-owned, satellite dish systems across North America had to be repointed at a cost of $100 each. The British Broadcasting Company's news program on Houston's KPFT radio station went silent, so the station turned to the Internet to gain access to the program instead. Today's story was about criminals in Bombay who launder their money through the movie industry, and were prone to kill the Director if the movie bombed at the box office. Meanwhile, Data Transmission Network Corporation lost service to its 160,000 subscribers, costing the company over $6 million. Many newspapers and wire services noted that this was the day that the Muzak died, because the Galaxy IV also took with it the feed from the Seattle-based music service. Many who previously thought that 'elevator music' was annoying, realized just how much they actually missed hearing it for the first times in their lives. We can mostly survive these kinds of annoyances, but the impact of the satellite outage spread into other life-critical corners of our society as well. Hospitals had trouble paging their doctors on Wednesday morning for emergency calls. Potential organ recipients who had come to rely on this electronic signaling system to alert them to a life-saving operation, did not get paged. In the ensuing weeks, many newspapers including USA Today wrote cautionary stories about how we have become too reliant on satellites for critical tasks and services in our society. Even President Clinton ordered a complete evaluation of our vulnerability to high-tech incidents, some of which could be caused by terrorists.

This outage was followed on June 13th by the loss of the primary control processor on the Galaxy VII, and an identical problem with the DBS-1 satellite on July 4. PanAmSat Corporation, the owner of the three Hughes model HS-601 satellites, was never able to identify a clear cause for these failures although they claimed that 'tin wiskers' might explain the sudden electrical outage due to a short circuit. International Space Industry Report carried a headline "Hughes Hit Hard by Satellite Failures". *The failures have sent Hughes scrambling for an explanation, and left industry analysts wondering whether other Hughes-built satellites of the same family may be subject to similar problems"*

Space physicist Dan Baker and his colleagues at the University of Colorado, however, uncovered satellite evidence for a very disruptive space environment spawned by the April-May solar activity episode. There was a major NASA, POLAR satellite anomaly on May 6, and more than a dozen anomalies plagued Japan's Global Meterological Satellite system between May 4-7.

The 2000 Bastille Day Storm

July 14, 2000 – The Bastille Day Flare or Bastille Day Event was a powerful solar flare on July 14, 2000, occurring near the peak of the solar maximum in solar cycle 23. Active region 9077 produced an X5.7-class flare, which caused an S3 radiation storm on Earth fifteen minutes later as energetic protons bombarded the ionosphere. It was the biggest solar radiation event since 1989. The proton event was four times more intense than any previously recorded since the launches of SOHO in 1995 and ACE in 1997. The flare was followed by a full-halo coronal mass ejection[1] and a geomagnetic super storm on July 15-16. The extreme level, G5, was peaked in late hours of July 15.

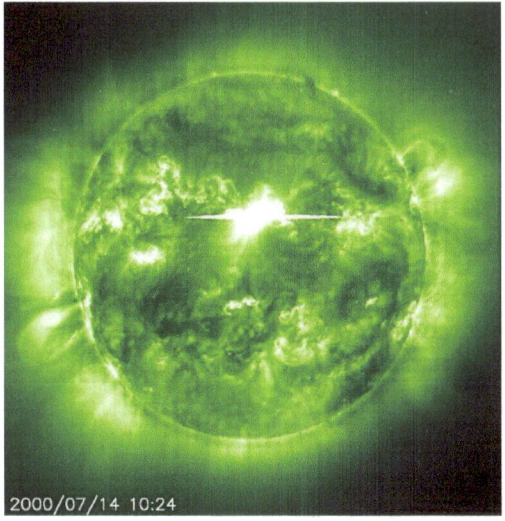

The "Bastille Day" solar flare as seen by SOHO's EIT instrument. (Credit: NASA)

July 17, 2000 - *Minor Damage Reported from Geomagnetic Storm* –A severe geomagnetic storm that hit Earth this weekend interfered with data from at least one weather satellite and some power systems, government scientists said today. But forecasters at the National Oceanic and Atmospheric Administration said the worst of Saturday's storm, caused by a huge solar flare on Friday, was over. 'It has died down now' said Craig Sechrest, an agency forecaster in Boulder, CO. The storm, which hit earth's magnetic field at 10:40 a.m. Saturday, was the biggest since October 1989 the agency's web site said. It was classified as severe to extreme with potential effects on radio communications. Power systems and aviation. The Federal Aviation Administration had no reports of problems, a spokeswoman said. Mr. Sechrest said that some power companies had minor problems and that any erroneous data from satellites was not enough to make the satellites unusable. [New York Times, July 17, p. A17]

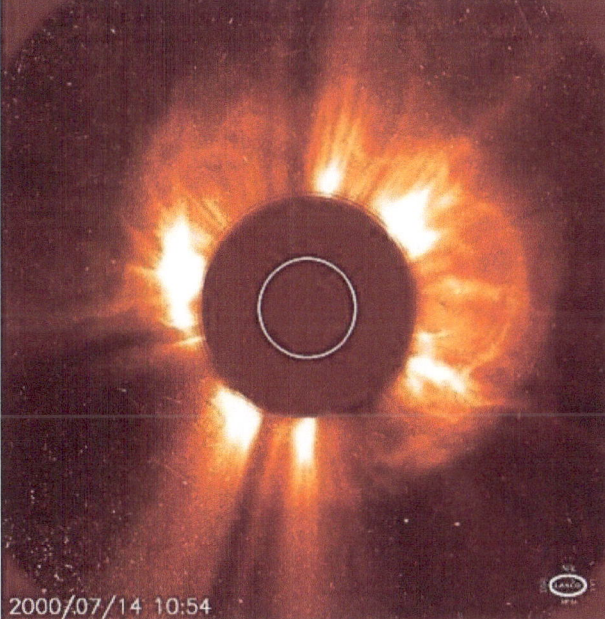

The CME launched by the Bastille Day Event forms a near-perfect halo of ejected plasma around the disk of the sun, indicating that it was aimed almost directly at Earth. Launched on July 14, the plasma reacher Earth three days later on July 17.

Auroral corona, March 24, 2001 (Courtesy Jan Curtis)

March 31, 2001 – This Easter Storm was tracked from cradle to grave by NASA and ESA satellites. The CME emerged from tangled magnetic conditions overlying one of the largest sunspot groups seen in several decades. A relatively dense and strongly magnetized interplanetary shock wave hit Earth's magnetosphere at 8 p.m. EST on March 30th. Strong geomagnetic activity, including mid-latitude auroras, are likely to follow. On March 29, 2001, the largest sunspot in ten years crossed the solar disk. The fast-growing spot, called AR9393, covered an area of the Sun equivalent to the total surface are of 13 Earths!

July 15, 2001 - - Solar flare threatens the earth with storm [New York Times, July 16, 2001 p. 21]. Minor damage reported from geomagnetic storm [New York Times, July 17, 2001 p. A17]

The GPS Era

Navigation by satellite is not a new technology, and was first introduced by the US Navy in 1960 with the orbiting of five Transit satellites. This system was replaced by the NAVSTAR-GPS system in the 1970s. The first commercial use of satellite-based global positioning systems came less than 1 year after the next generation, 24-satellite 'Block I-GPS' constellation had been deployed in 1994, when Oldsmobile offered the GuideStar navigation system for its high-end automobiles. The GPS satellites provided an L1 channel at 1575 MHz capable of 10-meter-scale precision, that in 1990 was 'selectively degraded' to 100-meter precision. In 1999, President Clinton ordered that selective availability be turned off, and on May 1, 2000 the modern era of non-military GPS was ushered-in. Since 2000, the commercial applications of GPS have enormously expanded to include, not only car navigation aids, but oil extraction, fiber optic cable deployment, civilian aviation, emergency services, and even expanding public cellphone services, called apps, to locate nearby stores, restaurants and even parking spaces in downtown Manhattan! A report by Berg Insight (2011) indicates that GPS-enabled mobile phones reached 300 million units in 2011, and is expected to reach nearly 1 billion units by 2015.

One of the first unclassified studies to quantitatively assess GPS behavior under solar storm conditions was conducted, inadvertently, by NOAA in 2001. They had set up a network of 70 GPS receivers from Alaska to Florida to test a new weather observation and climate monitoring system called the GPS-MET Demonstration Network. A major geomagnetic storm between March 30-31 caused significant changes in the GPS formal error, and was correlated with the published Kp index during the course of the event. [NOAA, 2001]. Since then, a variety of anomalous changes in GPS precision have been definitively traced to, and found to be correlated with, geomagnetic storms and solar flare events. This also means that systems that rely on GPS for high-precision positioning have almost routinely reported operational upsets of one kind or another. For example [NOAA, 2004]:

October 29, 2003 - the FAA's GPS-based Wide Area Augmentation System (WAAS) was severely affected. The ionosphere was so disturbed that the vertical error limit was exceeded, rendering WAAS unusable. The drillship GSF C.R. Luigs encountered significant differential GPS (DGPS) interruptions because of solar activity. These interruptions made the DGPS solutions unreliable. The drillship ended up using its acoustic array at the seabed as the primary solution for positioning when the DGPS solutions were affected by space weather.

December 6, 2006 - the largest solar radio burst ever recorded affected GPS receivers over the entire sunlit side of the Earth. There was a widespread loss of GPS in the mountain states region, specifically around the four corners region of New Mexico and Colorado. Several aircraft reported losing lock on GPS. This event was the first of its kind to be detected on the FAA, WAAS network.

Apart from changes in ionospheric propagation, we have the problem that, if the GPS signal cannot be detected by the ground station, and the minimum of 4 satellites is not detected, a position solution will not be available at any accuracy. This situation can arise if the GPS signal is actively blocked or jammed, or if the natural background radio noise level at the L1 and L2 frequencies is too high. This can easily happen during radio outbursts that accompany solar flare events. This happened the day after the December 5, 2006 solar flare, and was intensively studied by Kintner at Cornell, and presented at the Space Weather Enterprise Forum in Washington, DC on April 4, 2007 [NOAA, 2007].

The 2003 Halloween Storm

October 29, 2003 - This Halloween Storm spawned auroras that were seen over most of North America. Extensive satellite problems were reported, including the loss of the $450 million Midori-2 research satellite. Highly publicized in the news media. A huge solar storm has impacted the Earth, just over 19 hours after leaving the sun. This is one of the fastest solar storm in historic times, only beaten by the perfect solar storm in 1859 which spent an estimated 17 hours in transit. A few days later on November 4, 2003 one of the most powerful x-ray flares ever detected, swamped the sensors of dozens of satellites, causing satellite operations anomalies....but no aurora. Originally classified as an X28 flare, it was upgrade to X34 a month later. In all of its fury, it never became a white light flare such as the one observed by Carrington in 1859. Astronauts hid deep within the body of the International Space Station, but still reported radiation effects and ocular 'shooting stars'.

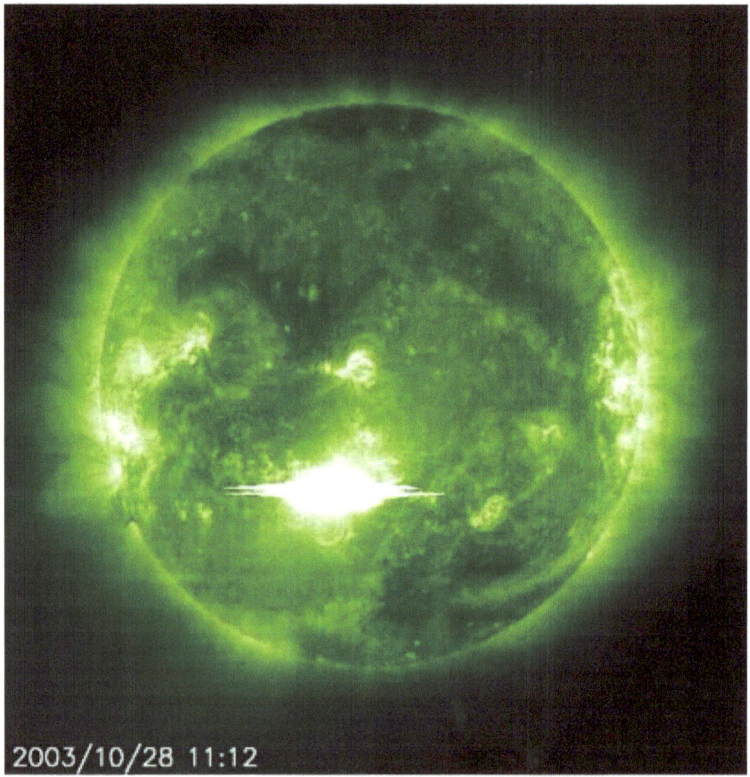

2003/10/28 11:12

The Halloween flare recorded by the SOHO spacecraft on October 28.

On October 23 American Michael Foale, Russia's Alexander Kaleri and Spain's Pedro Duque entered the space station after the autopilot docking of their spacecraft, two days after the Soyuz blasted off from Kazakhstan. Oct 28th - Meanwhile, satellite operators and power grid managers prepared to endure a potentially damaging event. And astronauts aboard the International Space Station took cover from heavier radiation sent out by the flare. They are not expected to be in any serious danger. Expedition 8 crew of Commander Mike Foale and Flight Engineer Alexander Kaleri will spend brief periods of time in the aft end of the Zvezda Service Module, which is the location aboard the Station most shielded from higher levels of radiation. The crew will spend about 20 minutes in Zvezda, twice on each orbit of the Earth for about three orbits, until the station phases out of the high radiation areas (high magnetic latitudes). Pedro Duque's diary from space: The curtains of the sky 24 October 2003:

Below your spacecraft the planet is black, almost like the sky. Like every night in orbit. Or maybe not? Far away, towards the horizon between the black of the sky and the black of the Earth, a yellow-green haze of irregular shape begins to show up. Too weird to be clouds. As you get closer it turns into more and more brilliant gauze veils, which form curtains coming up from the ground. Another turn of the spacecraft and you see them again, you are almost above them. The curtains get more and more defined - they are striped and reach very high, even higher than the spacecraft. You go straight towards them, and you can't avoid feeling somewhat uneasy - will this be dangerous? When you enter them, all the curtains next to you start to light up and change shape, as if you were walking among sunny window blinds. The phenomenon lasts for one minute, then two, even three, and nothing unusual can be felt, the spacecraft goes on with its very soft trip along the rail of its orbit. When you leave behind this phantasmagoric image and face the blackness again you feel relief, but also a sense of loss.

On October 29, Foale and Kaleri, the only humans currently outside the protection of Earth's atmosphere, retreated during peak exposure times to the living quarters of the station, which provided the best radiation protection. As a precaution, NASA shut down the station's robotic arm, which is the most exposed piece of hardware.

October 30, 2003 - Malmo Sweden, population 50,000 lost electrical power for 50 minutes [Pulkkinen et al., 2005]. "Since everything worked properly once power was restored, and since we have not had any further disruptions, a geomagnetic disruption is the likeliest

explanation," Sydkraft operations engineer Sven-Aake Andersson told Swedish news agency TT on Friday. The blackout was caused by the tripping of a 130 kV line. It resulted from the operation of a relay that had a higher sensitivity to the third harmonic (=150 Hz) than to the fundamental frequency (=50 Hz). The excessive amount of the third harmonics in the system has been concluded to have resulted from transformer saturation caused by GIC. Currents as high as 330 Amperes were recorded on the Simpevarp-1 transformer. [Wik et al. ,2009; SpaceDaily.com, 2003]

Damage to a transformer in South Africa.

October, 2003 - South Africa Transformer Damage. The ESKOM Network reported that 15 transformers were damaged by high GIC currents. Figure 6 shows one of the transformers in a view reminiscent of the legendary images of the 1989 Quebec transformer failure. [Murtagh, 2009]

November 4, 2003 – One of the most powerful x-ray flares ever detected, it swamped the sensors of dozens of satellites, causing satellite operations anomalies, but no aurora.

Originally classified as an X28 flare, it was upgrade by OAA scientists to X34 a month later. Astronauts hid deep within the body of the International Space Station, but still reported radiation effects and ocular 'shooting stars'.

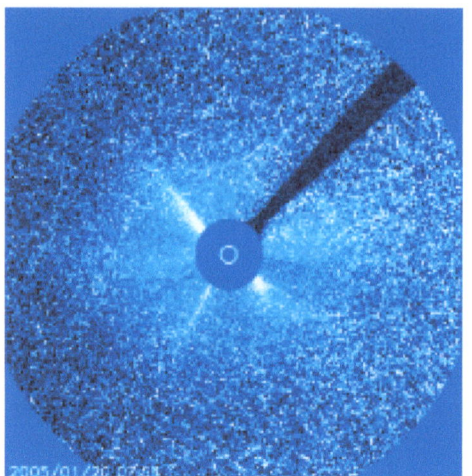

November 20, 2003 – No CME was involved in this storm, which appeared after Earth's passage through a high-speed coronal wind stream. Spectacular aurora were observed, and newspaper accounts abounded.

Solar particles blinding the imaging sensor on the SOHO spacecraft during the January 20, 2005 radiation storm.

The 2005 US Airlines Polar Detour Storm

January 20, 2005 – Largest SPE event in 50 years. A giant sunspot named "NOAA 720" exploded. The blast sparked an X-class solar flare, the most powerful kind, and hurled a billion-ton cloud of electrified gas (a coronal mass ejection, or CME) into space. Solar protons accelerated to nearly light speed by the explosion reached the Earth-moon system minutes after the flare; it was the beginning of a days-long "proton storm." An astronaut on the moon, caught outdoors on January 20, would have had almost no time to dash for shelter, and would have become sick. At first, he'd feel fine, but a few days later, symptoms of radiation sickness would appear: vomiting, fatigue, low blood counts. These symptoms might persist for days. The January 20 proton storm was by some measures the biggest since 1989. It was particularly rich in high-speed protons packing more than 100 million electron volts (100 MeV) of energy. Such protons can burrow through 11 centimeters of water. A thin-skinned spacesuit would have offered little resistance. Astronauts on the International Space Station (ISS), however, were safe. The ISS is heavily shielded, plus the station orbits Earth inside our planet's protective magnetic field. "The crew probably absorbed no more than 1 rem," said Francis Cucinotta, NASA's radiation health officer at the Johnson Space Center. On the moon, Cucinotta estimates, an astronaut protected by no more than a space suit would have absorbed about 50 rem of ionizing radiation. That's enough to cause radiation sickness. "But it would not have been fatal," he adds. [NASA press release, 11/8/2005]

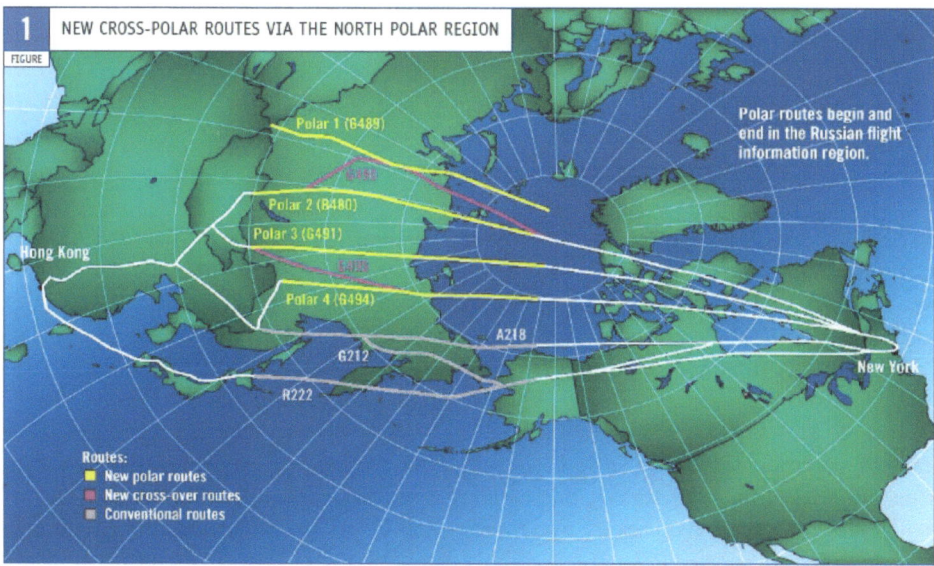

An example of polar routes used by passenger aircraft (Credit Boring)

Two major US airlines rerouted planes away from the polar areas to avoid additional radiation, said Bill Murtagh, a space weather forecaster for NOAA. Instruments on two of NOAA's satellites, SOHO and the Advanced Composition Explorer, were blinded by radiation contamination for several hours, and there are unconfirmed reports of problems on other satellites. The eruptions also required evasive action aboard the International Space Station. The two-man crew, Leroy Chiao and Salizhan Sharipov, ducked for cover inside the bulkier Russian side of the station when their orbit took them through the worst of the storm. High doses of radiation can cause health problems: astronauts are more prone to cataracts later in life, for example. [New Scientist, January 21, 2005]

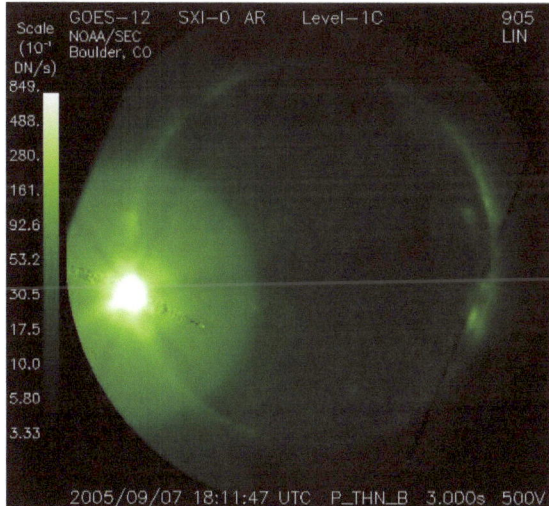

Image taken by the GOES X-ray imager shows the powerful X-ray flare in progress.

September 7, 2005 – A very large solar X-ray flare (X17) - The Sun unleashed a torrent of X-rays on Wednesday, blacking-out high-frequency radio communications in North and South America, including states around the Gulf of Mexico still recovering from Hurricane Katrina. And forecasters say more major solar flares may be on the way. The solar flare was the fourth largest in 15 years, in terms of X-ray emissions, and the fifth largest since 1976. The flare - which erupted at 1340 EDT (1840 GMT) - caused a complete blackout for those using high-frequency radio communications on the day-lit side of Earth. Air traffic controllers experienced moderate communications problems on Wednesday [New Scientist, 9/9/2005].

The Decline of Short-wave Broadcasting

Although telegraphic communication was the dominant victim of solar geomagnetic activity during the 1800s, by the mid-20th Century, virtually all telegraphic systems had been replaced by land-lines carrying telephonic communications, or by the rapid rise of short-wave broadcasting and submarine cables for trans-continental communication. At its peak in ca 1989, over 130 million weekly listeners tuned-in to the BBC's World Service. Once the Cold War ended, short-wave broadcasting and listening went into decline. Today, less than 1/3 of the stations on the air in 1970s are still operating. Compared to other forms of communication (e.g. web-based programming) shortwave is very expensive in terms of setting up a radio station, or providing operating costs to purchase megawatts of broadcasting power. Nevertheless, by December 2011 an estimated 33% of the human population had access to the Internet, and its vast network of formal and informal 'news' aggregators, including online versions of nearly all of the former shortwave broadcasting stations.

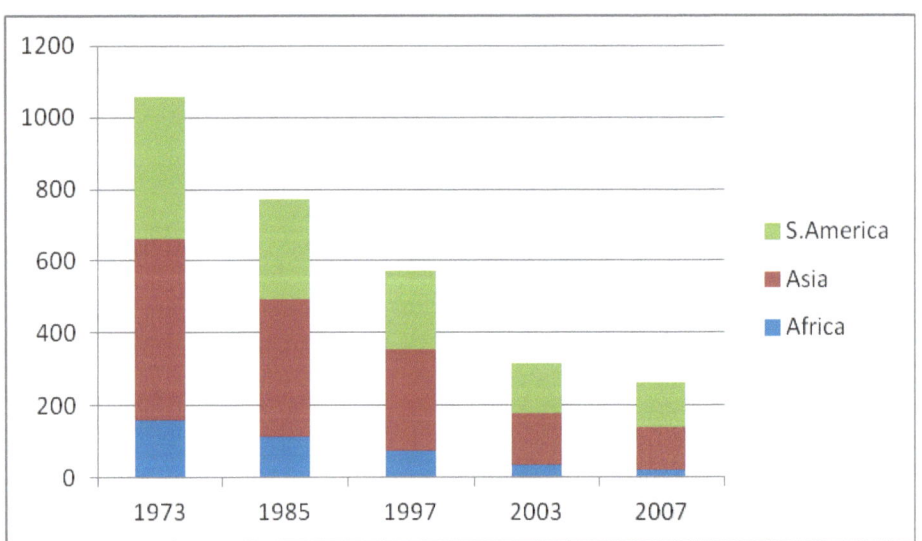

The number of short wave stations (vertical axis) has dropped dramatically since the advent of the World Wide Web and other wireless media, which now provide the main source of news reporting in the 21st century. (Data courtesy Careless, 2010)

Although shortwave broadcasting is a ghost of its former self, there are still a number of functions that it continues to serve in the 21st Century. It is a back-up medium for ship to shore radio, delivering state-supported propaganda to remote audiences, time signals (WWV), encrypted diplomatic messaging, rebel-controlled, clandestine stations, and the mysterious 'Numbers Stations'. There also continues to be a die-hard population of amateur radio 'hams' who continue to thrill at DXing a dwindling number of remote, low-power stations around the world when the ionospheric conditions are optimal. Sometimes, these Ham operators serve as the only communication resource for emergency operations. For example, during Hurricane Katrina in 2005, over 700 ham operators formed networks with local emergency services, and were the only medium for rapidly communicating life-saving messages. Despite the lack of public interest or awareness of the modern shortwave band, its disruption could leave many critical emergency services completely blind and unresponsive in a crisis.

Short wave broadcasting played such a key societal role during the first-half of the 20th century that millions of people were intimately familiar with its quality, program scheduling, and disruptions to this medium. Any disruption was carried as a front page story in even the most prestigious newspapers such as the New York Times. Although as we noted before, contemporary public contact with shortwave radio is nearly zero, today there are some places where SW is still in limited use, and where the public in those regions would be as conversant with SW fade-outs as the western world was in ca 1940. For instance, China is expanding its SW broadcasting to remote populations across China who do not as yet have access to other forms of communications networks. Even today, short wave outages still make the news.

Sunspot Cycle 24

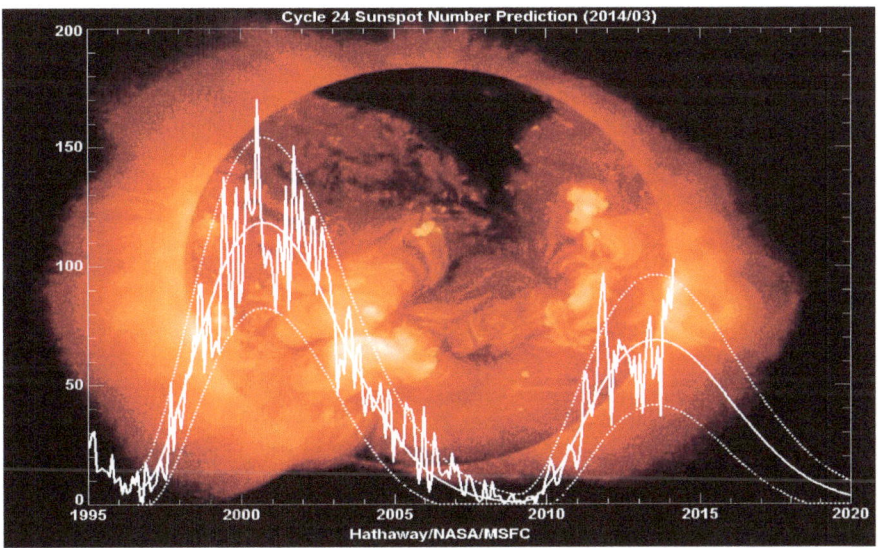

The figure above shows the average monthly sunspot numbers from 1995 at the start of Cycle 23, to the predicted end of the current sunspot cycle in ca 2020. The plotted curves give the maximum, average and minimum levels predicted by different methods. It is clear that we are close, or have just passed sunspot maximum for Cycle 24, and that our current cycle is about half the intensity of Cycle 23. This means fewer solar storms, but it does not guarantee that we will not have one or two major events in the remaining time to 2020. In fact, based on studies of previous sunspot cycles, the most intense storms happen within one to two years after sunspot maximum, so there is still time between 2014 and

2016 for The Big One. How big it will be is a statistical question, but storms ½ as intense as the 1859 Superstorm happen every 20-30 years so we may yet get a storm big enough to tell our grandchildren about!

In 2006, a number of prediction schemes were exercised in order to forecast what Cycle 24 might be like, but it was soon recognized from the steady pile up of 'spotless days' that Cycle 24 would be qualitatively different than any cycle since ca 1906. There was even some speculation that a new Maunder Minimum might be at hand. By 2008, a survey by solar physicist Dean Pesnell at NASA/Goddard showed that there were over 50 predictions for the peak year and sunspot numbers for Cycle 24 using many different, and often independent approaches, both statistics-based and physics-based (see Charbonneau, 2010). The earliest 'physics–based' predictions have suggested that Cycle 24 might be as severe as Cycle 23, and perhaps even worse than any cycle in the last 400 years. This turned out not to be in line with the trends that began to materialize by ca 2010 and now estimates have been revised downwards to a peak near 60 and a peak time during the second-half of 2013.

Recent studies using the GONG network had not detected the sun's torsional oscillation, which usually is an advanced indicator of east-west, subsurface plasma flows, and that have been identified as the harbingers of sunspot formation in the next cycle. Before Cycle 24 had begun, these same torsional oscillations had been detected and correctly predicted the late-arrival of the onset of Cycle 24. The concern is that the start of Cycle 25 may be significantly delayed to ca 2022, or possibly not occur at all [Penn and Livingston, 2006; Hill, 2011].

The 2010 Galaxy 15 Zombie Satellite Storm

April 5, 2010 - Galaxy 15 experienced an electrostatic discharge (ESD) that caused a severe malfunction, rendering the satellite capable of re-transmitting any received signal at full-power, but not able to receive new commanding [de Shelding, 2011]. Reports cited a space weather event on April 5 as the probable cause of the electrostatic discharge that was the likely triggering event, however although Intelsat acknowledged the ESD origin, they categorically refuted the space weather cause in the April 5 solar event, preferring to declare that the origin of the ESD was unknown. A consequence of this type of satellite failure is that Galaxy-15 was potentially able to interfere with other GEO satellites as it came within 0.5 degrees of their orbital slots. Thanks to careful, and complex, maneuvering of the satellites to maximize their distance from this satellite as it entered their orbital slots, AMC-11, Galaxy-13, Galaxy-18, Galaxy-23 and SatMex-6 and Anik F3 were able to reduce or eliminate interference, and no impacts to broadcasting were reported or acknowledged. "The fact that you haven't heard about channels lost or interference is the proof that we have been able to avoid issues operationally," said Nick Mitsis, an Intelsat spokesperson. "I don't want to underplay that." [Clark, 2010]. In January 2011 commanding of the satellite resumed and its 'zombisat' moniker has been changed to 'phoenix'.

The 2011 Chinese Shortwave Blackout Storm

August 5, 2011, the combined cloud of three consecutive CMEs produced brilliant auroras, reported as far south as Oklahoma and Alabama. The geomagnetic storm reached a G4 (severe) level, enough to make power outages. It was one of the strongest geomagnetic storms in years. In the southern hemisphere, auroras could have been seen as far north as South Africa, Southern Chile and Southern Australia. The CMEs were hurled by three M-class flares erupted in active sunspot 1261: M1.4 on August 2, M6.0 on August 3 and M9.3 on August 4. The major solar flare on August 9 caused fade-outs in the SW broadcasts of Radio Netherlands World, but after an hour, broadcasting had returned to its normal clarity. Solar flare disrupts RNW short wave reception [RNP, 2011]. This was the first major SW blackout in China since the X7.9-class flare on January 21, 2005, which affected Beijing and surrounding eastern population centers. [Xinhuanet, 2005]. On February 15, 2011 another large solar flare disrupted southern Chinese SW broadcasting. The China Meteorological Administration reported an X2.2-class flare at that time. [Xihuanet, 2011]. The January 23, 2012 M9-class solar flare disrupted broadcasts on the 6 – 20 meters bands across North America, and severely affected the UHF and VHF bands for a period of a few hours. [SWA, 2012]

Image taken on January 22, 2012 by the Solar Dynamics Observatory of the erupting class M8.7 X-ray solar flare (Credit: NASA/SDO)

August 25, 2011 South Africa's $13 million LEO satellite SumbandilaSat failed, and the explicit cause was stated publically to be 'damage from a recent solar storm', which caused the satellite's onboard computer to stop responding to commands from the ground station. This was not, however, the first time this satellite was damaged by radiation. Shortly after its launch in September 2009, radiation caused a power distribution failure that rendered the Z-axis and Y-axis wheel permanently inoperable, meaning that the craft tumbles as it orbits and has lost the ability to capture imagery from the green, blue and xanthophyll spectral bands. The reason given for the lack of proper radiation hardening was that there was not enough money to do this properly, and the satellite was built from commercial off-the-shelf (COTS) equipment. Moreover, SumbandilaSat was intended only as a technology demonstrator. [Martin, 2012]

The 2011 Anik F2 Satellite Storm

October 6, 2011 The Anik F2 experienced a 'technical anomaly' and entered a Safe Mode that caused it to stop functioning and turn away from Earth. The Boeing satellite was launched in 2004 and was expected to function for 15 years. The owner of the satellite, Telsat, indicated in public news articles that they did not believe the problem had to do with the arrival of a CME that reached Earth early the same morning, but was caused by some other unspecified internal issue with the satellite itself. It is the first serious anomaly of its kind since the satellite was launched in 2004. What the news reports failed to mention was that the sun has been relatively quiet for the majority of this 7 year period. [Mack, 2011]

The temporary outage of Anik F2 caused a number of problems that impacted millions of people covered by this satellite service. WildBlue satellite ISP in the United States uses Anik F2 to provide broadband services to about a third of its customers. A total of more than 420,000 subscribing households mostly in parts of rural America lost service for several days, along with ATM service. Canadian Broadcasting Corporation indicated that 39 rural communities, and 7,800 people lost long-distance phone service. The satellite is also used for air traffic control, causing the grounding of 48 First Air flights, and 1000 passengers, in northern Canada. Communities in the North West Territories were instructed to activate their emergency response committees, and start using their Iridium phones. [Mack, 2012; CBS News, 2012; Marowits, 2011]

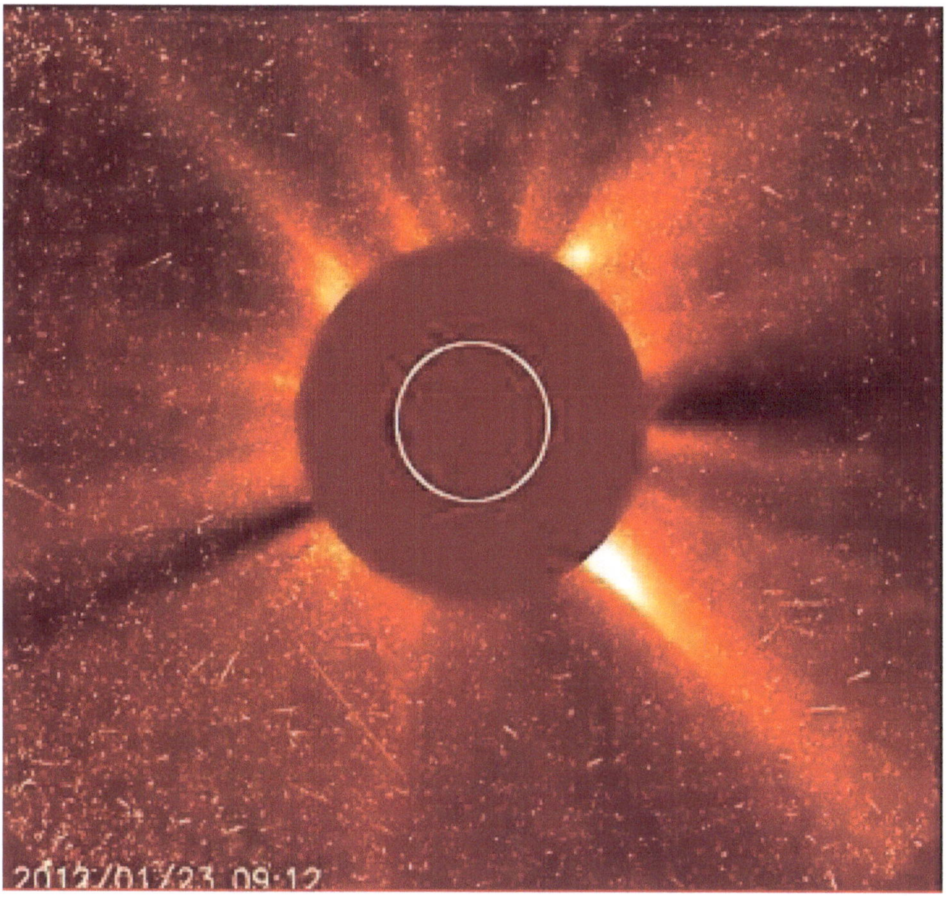

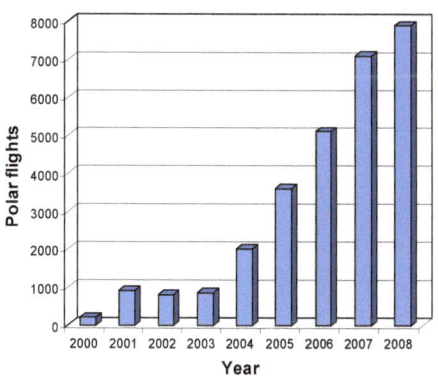

The energetic particles from the January 23, 2012 CME cause 'snow' in the sensors from the direct hit.

January 23, 2012 – The Coronal Mass Ejection caused by a January 22 M8.7 class flare, hit Earth at 1,400 miles per second, causing a geomagnetic storm and a radio blackout. The storm affected air traffic and may satellites. Many airliners have been avoiding the North Pole routes because they are more exposed to the proton storm, which disrupts High Frequency radio communications. HF data links are crucial to modern air flight, as they keep aircraft connected to Air Traffic Control. Due to the structure of the magnetic field that surrounds Earth, the polar regions have very little protection against outbursts of solar radiation, so any airplane crossing that area could be exposed to this mayhem.

The number of passengers flying 'polar routes' continues to sharply increase each year to a current level of nearly 1.7 million passengers each year [Murtagh, 2010]. A number of national and international studies have been conducted to assess the radiation exposure to passengers during active space weather conditions and under otherwise normal circumstances. Normal background radiation doses on the ground are typically 0.3 microSv/hr. For passengers and flight crews, the actual cabin exposure varies with the geographic latitude of the flight, the altitude of the flight, and the combined GCR and solar fluxes of particles. For example, Bottollier-Depois et al (2000) determined from direct measurements at maximum solar activity in 1991-1992 and at minimum activity in 1996-1998, the lowest mean dose rate measured was 3 microSv/hr during a Paris-Buenos Aires flight in 1991. The highest rates were 6.6 microSv/hr during a Paris-Tokyo flight on a Siberian route and 9.7 microSv/hr on Concorde in 1996-1997. A number of similar studies since then have supported the idea that there is in fact some additional passenger and flight crew radiation exposure caused by space weather, however the levels are cumulatively very low for the vast majority of passengers who travel infrequently during the year.

Nevertheless, some airlines that fly polar routes, such as United Airlines, are sensitive to solar storm events, not necessarily for the added radiation load, but for the disruption of emergency high-frequency communications with ground controller, which violates FAA safety regulations.

On June 4, 2013, the National Space Weather Program Council convened the fifth annual Space Weather Enterprise Forum. At the meeting, Thomas Fahley and Gregg Scott of Delta Airlines detailed how solar flares and radiation storms caused multiple flights to be redirected away from the poles during the January, 2012 flare event. To avoid communications blackouts and high-energy radiation, which are concentrated around the poles during solar storms, more than 16 transcontinental flights were detoured to more southerly latitudes. On a per-flight basis, the detours consumed as much as 9,950 extra pounds of fuel and added as much as $4,507 to the price tag of each flight. Delays and missed connections multiplied costs even more. And that was just Delta. Other major airlines around the world have similar stories to tell. [NASA press release; June 4, 2013]

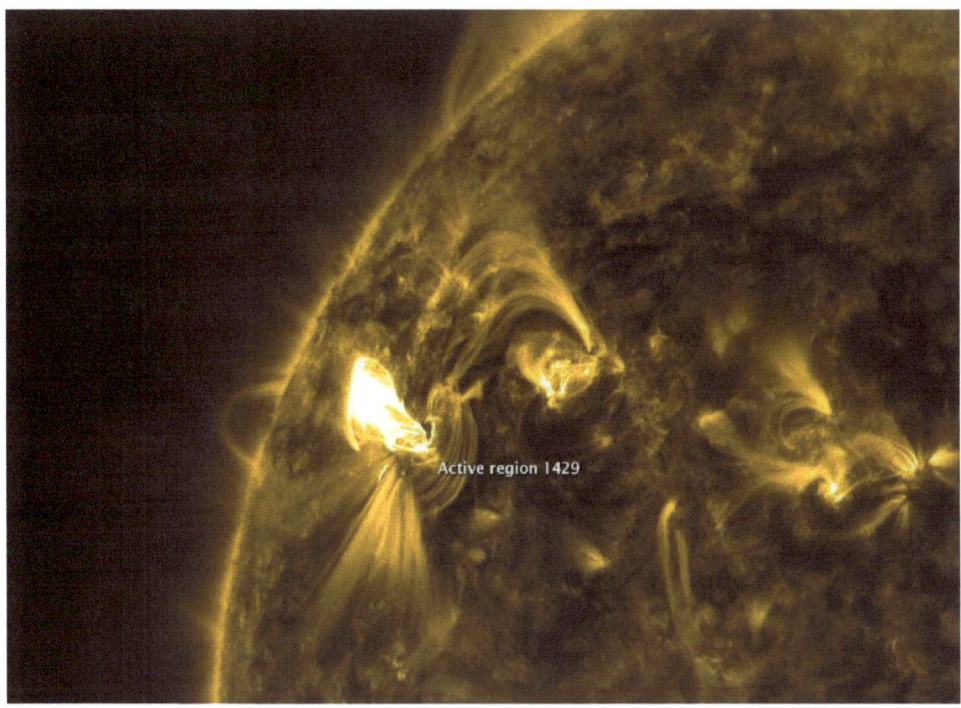

March 5, 2012 Following several minor C-class flares, M-class flares and CMEs registered in previous weeks and days, active region 1429 erupted an X1.1-class flare on March 5 at 0413 GMT (see image above by NASA's SDO satellite). The X-rays reached Earth in minutes, causing an R3 (strong) radio blackout over China, India and Australia, according to NOAA. Sunspot region 1429, whose size was the half of that of Jupiter, was being particularly active since it appeared on March 2. The CME arrived at the Earth on March 7 and caused a G2 (moderate) geomagnetic storm. Just hours after ejecting the X1.1-class flare, AR1429 produced several minor C and M-class flares in chain.

In late March, the US Air Force Space Command reported that the solar storms of March 7–10 could have temporarily knocked American military satellites offline. NASA also reported that these powerful flares heated Earth's upper atmosphere. From March 8 to March 10, the thermosphere absorbed 26 billion kWh of energy. Infrared radiation from carbon dioxide and nitric oxide, the two most efficient coolants in the thermosphere, re-radiated 95% of that total back into space.

There was also a report at the online newsletter gps.gov that the storm disrupted satellite communications and forced airlines to reroute some flights. But so far, no major GPS problems had been reported as a result of the event. The U.S. network of Continuously Operating Reference Stations (CORS), which monitors GPS daily from over 1,800 locations, observed only slight changes to GPS reception in some parts of Alaska on March 7 and 9.

July 23, 2012 – Although the spectacular CME from the sun on this date completely missed a collision with Earth, if the CME had erupted about one week earlier, Earth would have been squarely in the line of fire. According to data from the STEREO-A spacecraft (a solar observatory), the solar storm of July 23, 2012, was every bit as potent as the Carrington storms. If that CME had hit Earth, the resulting geomagnetic storm would have been comparable to the Carrington Event and twice as bad as the March 1989 Quebec blackout. This image from STEREO-A spacecraft. clocked this giant cloud, known as a coronal mass ejection, or CME, as traveling between 1,800 and 2,200 miles per second as it left the sun.

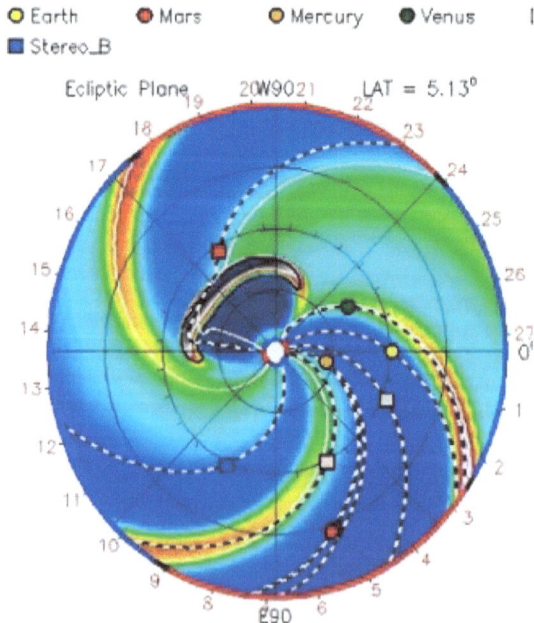

The Space Weather Research Center at the NASA Goddard Spaceflight Center, modeled the July 23, 2012 CME using a program called ENLIL. The figure shows the pinwheel shape of the solar wind out to the orbit of Mars (red dot). The location of Earth (yellow dot), Venus (green dot) and mercury (gold dot) are also shown to scale. The CME is the arc-shaped cloud approaching the STEREO A spacecraft (red square).

It turns out that the active region responsible for producing the July 2012 storm didn't launch just one CME into space, but many. Some of those CMEs "plowed the road" for the superstorm. The July 23rd CME was actually *two* CMEs separated by only 10 to 15 minutes. This double-CME traveled through a region of space that had been cleared out by yet another CME four days earlier. As a result, the storm clouds were not slowed down as much as usual by their transit through the interplanetary medium. The Carrington event was also associated with multiple eruptions, and this may turn out to be a key requirement for extreme events. Many researchers considered us to be very lucky that this storm was not headed for Earth, because all of the indicators suggested it would have been a Carrington-class event!

May 13, 2013 Solar activity increased rapidly in mid-May 2013 with four consecutive strong flares in two days. These powerful bursts all surged from the sunspot AR1748, located in the eastern limb of Sun. AR1748 emitted the first flare, an X1.7-class, on May 13, peaking at 02:17 UTC. This event was quickly followed the same day at 16:09 UTC by a X2.8-class flare (see above image by NASA's SDO satellite). On May 14 at 01:17 UTC the same sunspot emitted a X3.2-class flare, the third strongest of current solar cycle so far. This was followed by an X1.2-class flare at 01:52 UTC on May 15. The four X-ray bursts generated a R3 (strong) radio blackout in upper atmosphere. Every X-ray event was followed by a massive coronal mass ejection (CME). The first three CMEs were not directed toward Earth, but the fourth CME produced a G1 (minor) geomagnetic storm on May 18. A S1 (minor) proton storm event was also detected in connection with the May 15 X1.2 flare.

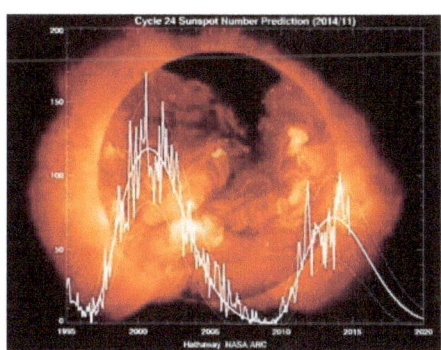

By the end of 2014, the sunspot counts are beginning their decline towards sunspot minimum from a peak near 2020.

Current cycle status ca December 2014. (Credit: David Hathaway, NASA/Marshall SFC) Current updates available at:

http://solarscience.msfc.nasa.gov/predict.shtml

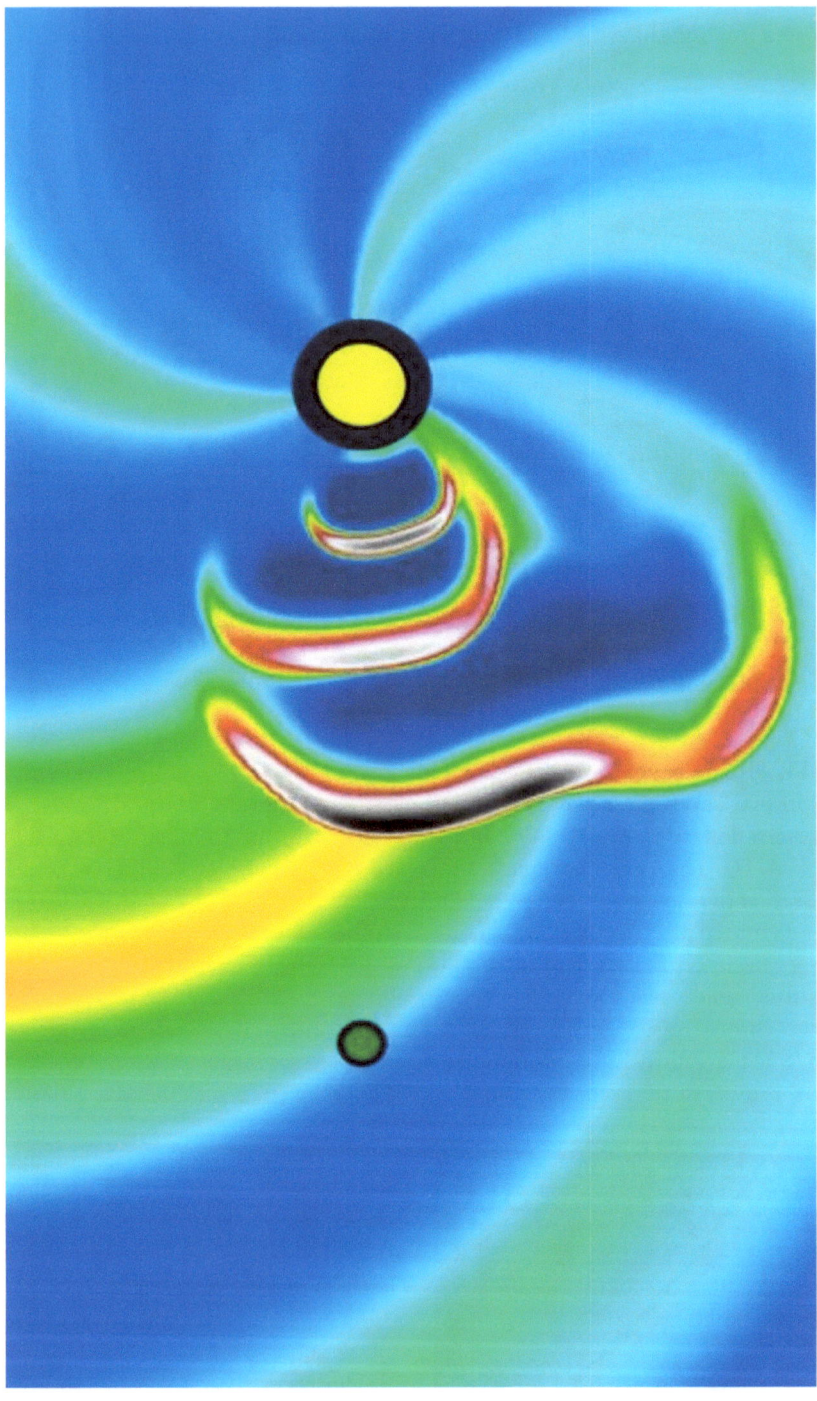

A physics-based NOAA space weather forecast model, which forecast the arrival of three successive coronal mass ejections in early August, 2011. The green dot is the location of Earth in its orbit at that time. The coloration indicates the density of the interplanetary plasma. https://www.youtube.com/watch?v=G9NQlGTFQIE

8 They call it *Space Weather*

It is an elementary exercise to follow the general trending of the last 23 sunspot cycles and predict that the current one will have a roughly-11 year period, followed no doubt by, Cycle 25, 26, etc. into the future. Of course we have historical evidence from the Maunder Minimum that, on occasion, the 11-year cycle seems to vanish entirely. A diminished solar activity cycle means that there are proportionately fewer CMEs, flares and SPEs to worry about, and so one expects a significantly lower risk of GICs in the electric power grid, satellite anomalies and a lowered risk for excessive radiation exposure for astronauts. Since the flux of cosmic rays is anti-correlated with the solar activity cycle, this still means that we will experience enhanced cosmic rays fluxes.

Compared to terrestrial weather, space weather as we have experienced it during the Space Age, is in the 'noise' in terms of economic impact. The difference between one or two extreme weather events in 2011, out of the 37 recorded world-wide, is about $2 billion. This equals the economic impact of ALL the commercial satellites lost in the entire 23rd sunspot cycle 11-year solar cycle ending 2006!

Cycle 23 will no doubt be seen by historians as a watershed moment in space weather history. Prior to Cycle 23, there was little or no public discussion about space weather vulnerability during the Space Age, although our grandparents surely knew all about the practical consequences of space weather and the insufferable short wave outages. With Cycle 23, we had SOHO providing the public with dazzling and ominous movies of solar storms, and many popularizers, including myself, who went on the stump to sort out for the public all the ways in which we could be affected. Then, just before the famous Halloween Storm of 2003, we had the first high-profile Congressional hearing about space weather in the context of why NOAA's Space Environment Center (SEC) budget should not be halved. Once Homeland Security became involved, we then had a new round of hearings about our infrastructure vulnerability to space weather events. The Space Weather Forum was held in Washington, DC on Capitol Hill in June 2008 to educate Capitol Hill about space weather issues. Meanwhile, researchers began the difficult task of trying to quantify what these impacts could cost us and the social disruption that might follow. Here is a short summary of some key documents and research findings that have transformed Space Weather studies into a genuine and well-recognized national security issue by 2015.

August 1988, Oak Ridge National Laboratory and the National Research Council published '*Evaluation of the Reliability for the Offsite Power Supply as a Contributor to the Risk of*

Nuclear Plants'. This set the stage for considering the impact of space weather-related GICs on the reliability and safety of nuclear power plants (Kirby et al., 1988).

April 1989, Northwest Power Coordinating Council (NPCC) approved the document *"Procedures for Solar Magnetic Disturbances Which Affect Electric Power Systems"* which has been updated several time.(NPCC, 1989)

Kappenman (1997) provides an extensive record of modeling the US power grid with increasingly more sophisticated models of the electrodynamics of GICs and exhaustive studies of the North American electric grid network at the component level. His efforts use historical geomagnetic storms (e.g. 1921 event) and extrapolate their impact on the contemporary electric power grid. Among the forecasts are for year-long recovery periods costing over $1 trillion in GDP.

Teisberg and Weiher (2000) estimated that the economic benefits of providing reliable warnings of geomagnetic storms to the electric power industry (alone) would be approximately $450 million over three years (note that this doesn't include any other impacted industries). This is well above the $100 million cost of a new operational satellite that would provide such warnings (ACE, Triana)

October 2003 – *'What is Space weather and who should forecast it?'* Congressional Hearing on Space Weather held before the Subcommittee on Environment, Technology, and Standards, Committee on Science, House of Representatives, One Hundred Eighth Congress, first session, October 30, 2003, (Congress, 2003)

December 2005, Idaho National Laboratory and the National Research Council published *'Reevaluation of Station Blackout Risk at Nuclear Power Plants--Analysis of Station Blackout Risk.'* The executive summary from this report reads in part: "The availability of alternating current (ac) power is essential for safe operations and accident recovery at commercial nuclear power plants." (INL, 2005)

Odenwald and Green (2007) modeled the economic losses to commercial satellites in LEO, MEO and GEO orbits and deduced that an 1859-scale 'superstorm' arriving near sunspot maximum could cost $50 billion in lost revenue and assets. A summary of this research appeared in the August 2008 issue of *Scientific American* magazine.

April, 2008: *"Report of the Commission to Assess the Threat to the United States from Electromagnetic Pulse (EMP) Attack: Critical Infrastructures"*. The US Congress funded a vulnerability assessment research under the National Defense Authorization Act to evaluate the impact of an electromagnetic pulse (EMP) from a high altitude nuclear detonation by a terrorist event on the nation's critical infrastructure including the electric grid. The same study also discussed geomagnetically-induced currents. (EMP Commission, 2008)

2008 *'Severe Space Weather Events—Understanding Societal and Economic Impacts Workshop Report'*. The National Academy of Sciences determined that severe geomagnetic storms have the potential to cause long-duration outages to widespread areas of the North American grid. (NAS, 2008)

June 2010, "*High-Impact, Low-Frequency Event Risk to the North American Bulk Power System*," jointly sponsored by NERC and the Department of Energy, NERC now concedes that the North American power grids have significant reliability issues in regard to High-Impact, Low-Frequency events such as severe space weather. The NERC report explains commercial grid vulnerability to space weather (NERC, 2010)

October 2010, '*Electromagnetic Pulse: Effects on the U.S. Power Grid*', Oak Ridge National Laboratory released a series of comprehensive technical reports for the Federal Energy Regulatory Commission (FERC) in joint sponsorship with the Department of Energy and the Department of Homeland Security. These reports disclose that the commercial power grids in two large areas of the continental United States are vulnerable to severe space weather. The reports conclude that solar activity and resulting large earthbound CME, occurring on average once every one hundred years, would induce a geomagnetic disturbance and cause probable collapse of the commercial grid in these vulnerable areas. The replacement lead time for extra high voltage transformers is approximately 1-2 years. As a result, about two-thirds of nuclear power plants and their associated spent fuel pools would likely be without commercial grid power for a period of 1-2 years. (Oak Ridge Labs, 2010)

Armed with all this bad news, it has become commonplace for Reporters to quote these studies and offer titles such as 'A big solar storm could cost $2 trillion, could be a global Katrina' or 'Solar storm buffets Earth: How protected is the US power grid?'. The danger is that, through constant repetition of this Doomsday theme, the public will become inured to the message in the face of the inevitable false alarms such as the January 2012 storm. While it is certainly important to keep the preparation message alive given the consequences to our infrastructure, as scientists and space weather forecasters, we need to be more careful with delivering this complex message to a Public increasingly eager for a yes or no answer to their safety.

Chief space weather forecasters at NASA/Goddard Spaceflight Center Yihua Zheng and Antti Pulkkinen (Credit: NASA/Chris Gunn)

Epilog

The last 200 years in reporting human and space weather impacts has passed through many stages and fads as new technological problems revealed themselves, and old ideas passed out of scientific fashion. In 2007 I completed a study of the history of space weather reporting in the print media (Odenwald, 2007) and showed that earlier accounts in the newspapers were more inclined to report problems because the impacts directly affected how news stories, themselves, were circulated (telegraph, wireless, teletype). During the 1950, post-war period, there are a broad array of communications media available to transmit and receive news stories, that any given space weather event causes little interruption in the flow of information, therefore, the obvious impacts are more subtle and difficult to apprehend. When this is combined with the lack of timely information on satellite, power grid, or radio anomalies from institutions locked in intense competitive struggles and attempting to demonstrate high reliability, the present dearth in impact reporting is understandable.

Since the 1960's, we have come to expect that news stories in the print media focus almost entirely on what the aurora look like and where they are sighted, revealing virtually nothing about the actual impacts that the storms had upon humans. Sometimes a critical story is not even considered newsworthy at all. One of the biggest space weather impacts, the Quebec Blackout of March 14, 1989, which affected 3 million people and cost $2 billion in Gross Domestic Product (as much as a major tornado) was only mentioned in two major metropolitan newspapers, The Boston Globe (March 14, 1989, p.6), and the Montreal Gazette (March 12-16, 1989). The aurora borealis itself was, however, mentioned in the San Francisco Chronicle (March 15, p.A3), the Los Angeles Times (March 14, p.2), the Philadelphia Inquirer (March 14, p.17A), the Baltimore Sun (March 13, p.3A, March 14, p.8B, March 15, p.4A), the Fairbanks Daily News (March 12, p.E10, March 14, p.3, March 15, p.1). According to the report in the Fairbanks Daily News ,"NBC News in Los Angeles asked to use five minutes worth of film (from a color movie of the aurora taken from Poker Flats) on the Nightly News Monday, but (Niel Brown, Director of the Poker Flats Research Range) doesn't know if the film was shown on national television". One obvious question to ask is whether the current level of space weather reporting in major newspapers has always been as we know it today, or whether previous decades and centuries were more enthusiastic in their reporting of solar storm events.

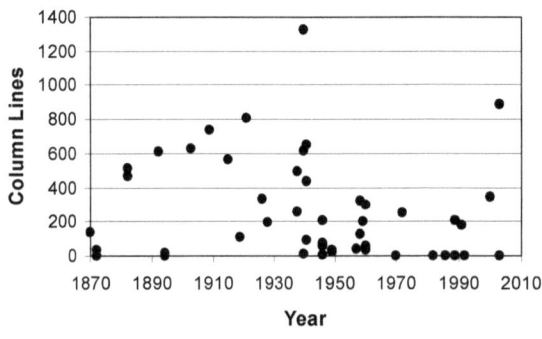

The graph shows that the total number of annual story lines published by the five major US newspapers showed a sharp decline in coverage after about 1950.

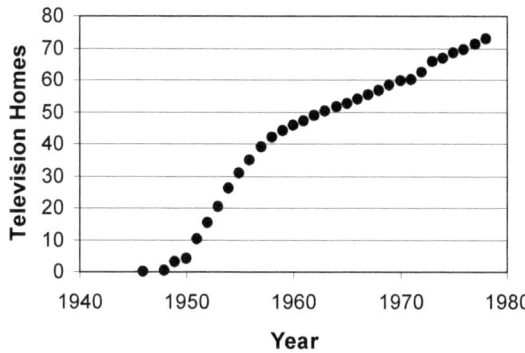

This graph shows that the growth of television sets between 1940-1960 surged after 1950, when an abrupt change in space weather story quantity and quality was identified.

The modern equivalent of the 19th century news media is far more complex and diversified across many modalities beyond newsprint. One of the biggest competitors to newsprint that entered the arena in the post-war years was television. The growth in the number of homes that owned at least one television, underwent a substantial change during the period from 1945-1955 when the largest change is seen in the publication of space weather events in the print media.

According to data from the Newspaper Association of America (2007), during the period after ca 1970, there has been a steady decline in regular newspaper readership. Surveys indicate that, today, about 25% of young Americans (ages < 25) and nearly 50% of older Americans (ages > 25), rely on TV news reportage as their primary source of information, rather than newsprint. Radio news reports also eclipse print as a secondary way to get news information, with 40% of young people and 60% of older people favoring radio news broadcasts on a daily basis. (Olander, 2003). To avoid competing with the evening news services, most newspapers bias their reporting to fill the morning editions. This growth of the morning paper has continued for nearly 30 years. Approximately six morning copies are sold for every evening edition (Edmonds, 2004).

The impact that these changes have had on the reporting of phenomena only of interest to night time aurora observers in the United States is that space weather stories may be more often placed in the evening editions of newspapers where the competition is relatively intense to attract the sagging numbers of readers who purchase the evening editions. More dramatic stories are selected to engage readers attention, at the expense of in-depth coverage of the comparatively benign impacts of space weather events. Meanwhile, it is not clear that the rise in television news coverage has taken up the slack in reporting space weather events during the immediate post-war period. The relevant, archival, data are not available to assess this point.

The second factor that may explain the lost ground in space weather news in the post-war era is that, according to Leonard David, Senior Space Writer for Space.com, increasingly, reporters need to be fed information in order to write a story. Most reporters do not have either the time or inclination to dig for the facts themselves. If a press release is not available on a particular science story, a reporter has to have a strong motivation to write the story from scratch. That motivation usually requires that there be some compelling human impact, disaster, or scientific angle involved that is immediately apparent, and could be generally understood and appreciated by the public. The Halloween Storm of October 29, 2003 was supported by press releases from NASA, and pro-active work by the NASA Public Affairs Office, and NOAA's Space

Environment Center, through their e-mail distribution networks that reaches over 1000 news reporters. Yet despite this Herculean effort to get the story out to the news media, the actual number of column-lines that resulted, about 888, is only comparable to similar severe storms reported between 1870-1950. Moreover, unlike the earlier newspaper stories, the modern-day stories did not report on specific impacts, but focused primarily on the more scientific elements of the phenomenon - consistent with the content provided by the NASA and NOAA/SEC press releases.

A reporter also needs to be savvy enough to call science contacts that can provide new information, in a timely manner, for a novel story that has to be written under a deadline. This runs into the predictable problem in space weather in which most sources inside NASA, the Department of Defense, the national power industry, or commercial satellite owners, do not want to talk about their various problems. Without access to actual stories of significant impacts, there is literally no other story that can be written other than purely descriptive, and far less compelling, accounts from eyewitnesses. Modern-day technological impacts are vastly under-reported compared to those 50-100 years ago because there are fewer commercial and government sources now willing to admit their vulnerability to the public.

Even when a story is well-written and compelling, there is a final hurdle to be surmounted. The more familiar the sun has become to the public, thanks to a constant stream of real-time imagery from NASA, the more familiar it has become to Editors, and therefore the less compelling. "Didn't we do a Sun story last week?"

Earlier accounts in the newspapers were more inclined to report problems because the impacts directly affected how news stories, themselves, were circulated (telegraph, wireless, teletype). Also, the public impact was greater and harder to camouflage as telegraph and wireless systems failed for significant fractions of the day. In fact, the vast majority of the accounts involved the very equipment that reporters used to gather and transmit their news reports. During the post-war period, there are a broad array of communications media available to transmit and receive news stories, that any given space weather event causes little interruption in the flow of information, therefore, the obvious impacts are more subtle and difficult to apprehend. When this is coupled with the lack of timely information on satellite, power grid, or radio anomalies from institutions locked in intense competitive struggles, and attempting to demonstrate high reliability, the present dearth in impact reporting is understandable.

Yet, considering that there are far more technological connections to space weather conditions today than there were 50 years ago, it is puzzling that the 'Golden Years' of space weather reportage has indeed passed, and the mediocre reporting of today is almost universally considered normal.

Bibliography

Amin, Massoud: "Powering the 21st Century: We can -- and must -- modernize the grid"; Today's Engineer: Mar 2005. http://www.todaysengineer.org/2005/Mar/grid.asp

Patterson, T., 2010, 'US Electricity Blackouts Skyrocketing', October 15, 2010, http://www.cnn.com/2010/TECH/innovation/08/09/smart.grid/index.html, accessed April 20, 2012.

Baker, D. 2000 "The occurrence of operational anomalies in spacecraft and their relationship to space weather", IEEE Transactions on Plasma Science, v. 28, p. 6.

Barcelona Journal, 1999, No. 49, http://www.publicacions.bcn.es/bmm/49/ang_12.htm

Bedingfield ,K. L., Leach, R.D., and Alexander, M.B. ,1996, "Spacecraft System Failures and Anomalies Attributed to the Natural Space Environment", NASA Reference Publication 1390, August 1996.

Berg, Jerome, 2008, 'Broadcasting On The Short Waves, 1945 To Today', (McFarland, NY), p. 22.

Berg Insight, 2009, '295 million GPS Handsets Sold in 2010; 940 million Expected in 2015 ', April 19, 2011, http://www.gpsbusinessnews.com/tags/Berg%20Insight/, accessed April 20, 2012.

Berkowitz, B., 2011, 'Extreme weather batters the insurance industry', February 09, 2011, http://www.reuters.com/article/2011/02/09/us-insurance-climate-idUSTRE7182XG20110209?pageNumber=1, accessed May 4, 2012.

Bolduc, L. 2002, "GIC Observations and studies in the Hydro-Quebec Power System", J. Atmos. Solar-Terr. Phys., v. 64, No. 16, pp. 1793-1802.

Bone, Neil, 1991, The Aurora: Sun-Earth Interactions, John Wiley and Sons (New York, 1991).

Boston.com, 2012, 'Computer glitch hits Brazil's biggest airline', March 02, 2012, http://articles.boston.com/2012-03-02/news/31117691_1_computer-glitch-check-in-system-airports, Accessed April 15, 2012.

Bottollier-Depois JF, Chau Q, Bouisset P, Kerlau G, Plawinski L, Lebaron-Jacobs L., 2000, 'Assessing exposure to cosmic radiation during long-haul flights.', Radiation Research, v. 153, pp. 526-532.

Brekke, Asgeir and Egeland, Alv, 1994, Nordlyset: kulturarv og vitenskap, (Grondahl and Dreyers Forlag AS, Oslo)

Browning, Orville Hickman, 1859, Diaries of Orville Browning for August 24-29 available at the Library of Congress, Manuscript and Rare Books Division.

Careless, James, 2011, 'Changes Continue for Shortwave', 10.26.2011, http://www.rwonline.com/article/changes-continue-for-shortwave/24684, accessed on April 15, 2012.

Careless, James, 2010, 'Whatever Happened to Shortwave Radio? ', 03.08.10, http://www.rwonline.com/article/whatever-happened-to-shortwave-radio/2842, Accessed on April 19, 2012.

CBS News ,2012, ' Satellite problems ground Nunavut flights', October 6, 2011, http://www.cbc.ca/news/technology/story/2011/10/06/north-satellite-phone-outage.html?cmp=rss, accessed April 15, 2012.

Charbonneau, P., 2010, 'Modeling solar and stellar dynamos', in 'Heliophysics: Evolving solar activity and the climates of space and earth', Carolus Schrijver and George Siscoe, eds., (Cambridge University Press; United Kingdom), pp.141-177.

Clark, S, 2010, 'Galaxy 15 'zombiesat' still alive after expected off date ',September 15, 2010, http://www.spaceflightnow.com/news/n1009/15galaxy15/, accessed April 21, 2012.

Colak, T. and Qahwaji, R. ,2007, "Automated Prediction of Solar Flares Using Neural Networks and Sunspots Associations," Advances in Soft Computing, DOI 10.1007/978-3-540-70706-6, Springer, 39,pp. 316-324, 2007. http://spaceweather.inf.brad.ac.uk/journals/wsc11_tc_rq.pdf, accessed April 20, 2012.

Congress, 2003, What is space weather and who should forecast it?: hearing before the Subcommittee on Environment, Technology, and Standards, Committee on Science, House of Representatives, One Hundred Eighth Congress, first session, October 30, 2003, Volume 4, http://www.solarstorms.org/CongressSW.html, accessed, April 23, 2012.

De Shelding, P. 2011, 'Electrostatic discharge crippled Galaxy 15 Intelsat says', January 13, 2011, http://www.spacenews.com/satellite_telecom/110113-electrostatic-discharge-galaxy15.html, accessed April 15, 2012.

Department of Energy: "Gridworks"; Office of Electricity & Energy Reliability: 2005. http://www.energetics.com/gridworks/index.html

Doherty, P.H., 2011, 'Space Weather Effects on GPS and WAAS', August 5, 2011, http://www.bc.edu/research/isr/spaceweathereffects.html, accessed April 15, 2012.

Dorman, L. I. et al., 2005, "Space weather and space anomalies", Annales Geophysicae ,V. 23, pp. 3009-3018.

Duncan, G., 2010, 'Use a cellphone? Join the rest of mankind!', http://www.digitaltrends.com/mobile/two-thirds-of-all-humans-use-a-cell-phone/, accessed May 2, 2012.

Eaton Corporation, 2011, 'Blackout Tracker: 2011 Annual Report', http://powerquality.eaton.com/info/GenOutput.asp?Quest_user_id=657593&leadG_Q_QRequired=False&menu=, April 20, 2012.

Editors, The great auroral exhibition of August 28th to September 4, 1859,

-----------, 1st article, Am. J. Sci., vol. 28, No. 84, 385–408, 1859.

-----------, 2nd article, Am. J. Sci., Second Series, vol. 29, No. 85, 92–97, 1860a.

-----------, 3rd article, Am. J. Sci., Second Series, vol. 29, No. 86, 249–265,1860b.

-----------, 4th article, Am. J. Sci., Second Series, vol. 29, No. 87, 386-397, 1860c.

Edmonds, 2004: The State of the News media', www.journalism.org

Electromagnetic Pulse Commission, 2008, 'Report of the Commission to Assess the Threat to the United States from Electromagnetic Pulse (EMP) Attack', April 2008, http://www.empcommission.org/docs/empc_exec_rpt.pdf, accessed April 20, 2012.

Energy Information Agency (EIA): "Electric Power Annual with data for 2005"; EIA/DOE: revised 09.11.2006. http://www.eia.doe.gov/cneaf/electricity/epa/epa_sum.html

Futron ,2002, "Satellite Insurance Rates on the Rise - Market correction or overreaction?", July 10, 2002.

Gallagher, P. T., Moon, Y.-J., & Wang, H. 2002, Sol. Phys., 209, 171.

Gellibrand ,1634, "A Discourse mathematical on the Variation of the Magnetical Needle together with its admirable Diminution lately discovered", (London, June 12, 1634).

Guillemin, Amedee, 1872 in 'The Heavens - An Illustrated Handbook of Popular Astronomy',(Richard Bentley & Son, London), The sunspots were observed by a Captain Davis on 30th August, 1839.

Hamachi-LaCommare, K. and Eto, J., 2004, 'Understanding the Cost of Power Interruptions to U.S. Electricity Consumers', http://certs.lbl.gov/pdf/55718.pdf, accessed April 21, 2012.

Hannah, Eric, 2004, 'Cosmic ray detectors for integrated circuit chips', US Patent 7,309,866; http://patft.uspto.gov/netacgi/nph-Parser?Sect1=PTO1&Sect2=HITOFF&d=PALL&p=1&u=%2Fnetahtml%2FPTO%2Fsr chnum.htm&r=1&f=G&l=50&s1=7,309,866.PN.&OS=PN/7,309,866&RS=PN/7,309,86 6. Accessed April 15, 2012.

Harris, Amelia Ryerse, 1859, Diaries of Amelia Ryerse Harris for August 22 and 28 available at the Library of Congress, Manuscript and Rare Books Division.

Idaho National Laboratory, 2005, 'Reevaluation of Station Blackout Risk at Nuclear Power Plants--Analysis of Station Blackout Risk', December 2005, http://www.nrc.gov/reading-rm/doc-collections/nuregs/contract/cr6890/, accessed April 20, 2012.

Kappenman, J., 2012, 'Impact of Severe Solar Flares, Nuclear EMP and Intentional EMI on Electric Grids', The Electric Infrastructure Security Summit, Third Annual World

Summit, London, 'http://www.eissummit.com/images/upload/conf/media/EIS_Kappenman_Part1.pdf', accessed April 20, 2012.

------------------, 2010, 'Low-frequency protection concepts for the electric power grid: Geomagnetically Induced Current and E3 HEMP mitigation', January, 2010, http://www.ferc.gov/industries/electric/indus-act/reliability/cybersecurity/ferc_meta-r-322.pdf, accessed April 20, 2012.

------------------, 2004 ,"Space Weather and the Variability of Electric Power Grids", in "Effects of Space Weather on Technology Infrastructure", I .A. Daglis, Ed.., (Kluwer: NY), pp. 257-286).

------------------, 1997, 'Geomagnetic storm forecasts and the power industry', *EOS Transactions*. January 29, 1997 p. 37.

Koons, H.C., Mazur ,J.E., Selsnick ,R.S., Blacke, J.B. ,Fennell, J.F. ,Roeder, J.L, and Anderson, P.C. ,1999, "The Impact of the Space Environment on Space Systems", Aerospace technical Report, TR-99(1670)-1, July 20, 1999. Space and Missle Systems Center, Air Force Materials Command ,2430 E. El Segundo Boulevard, Los Angeles Air Force Base, CA, 90245.

Lanzerotti ,L..J., Gary, D.E., Nita, G.M., Thomson, D.J. and Maclennan, C.G, 2005, 'Noise in wireless systems from solar radio bursts', Advances I nSpace Research, Vol. 36, No. 12, pp.2253-2257. http://solar.njit.edu/preprints/gary1218.pdf, accessed April 18, 2012.

Lincoln, Mary, 1959, Diaries of Mary Lincoln for August 28 and June 26 available at the Library of Congress, Manuscript and Rare Books Division.

Loomis, E. The great auroral exhibition of August 28th to September 4, 1859

------------, 5th article. Am. J. Sci., Second Series vol. 30 (88), 79–94, 1860a.

------------, 6th article. Am. J. Sci., Second Series 30 (90), 339–361, 1860b.

------------, 7th article. Am. J. Sci., Second Series 32 (94), 71–84, 1861a.

------------, 8th article, On the great auroral exhibition of August 28th to September 4, 1859, and on auroras generally. Am. J. Sci. 32 (96), 318–331, 1861b.

Mack, Eric, 2011, 'Major satellite outage affecting ISPs, ATMs and flights', October 6, 2011, http://news.cnet.com/8301-1023_3-20116846-93/major-satellite-outage-affecting-isps-atms-flights/, accessed April 15, 2012.

Marowits, Ross, 2011, ' Satellite outage disrupts communications in Canada's North, October 6, 2011, http://www.jocosarblog.org/jocosarblog/2011/10/satellite-outage-disrupts-communications-in-canadas-north.html, accessed April 15, 2012.

Martin, Guy, 2012, 'SumbandilaSat beyond repair', January 25, 2012, http://www.defenceweb.co.za/index.php?option=com_content&view=article&id=22870:

sumbandilasat-beyond-repair&catid=90:science-a-technology&Itemid=204, Accessed April 20, 2012.

Murtagh ,W., 2009, 'Space Weather; The next frontier in Severe Impacts', April 2, 2009, http://www.ametsoc.org/atmospolicy/documents/Bill_Murtagh_SpWx.pdf, accessed April 15, 2012.

Murtagh, B., 2010, 'Space weather impacts on aviation systems', September 8, 2010, http://www.easa.europa.eu/conferences/iascc/doc/Workshop%201%20Presentations/Workshop1_DAY%201/3_Murtagh_NOAA/Space%20Weather%20Impacts%20on%20Aviation%20Systems.pdf, accessed April 15, 2012.

National Academy of Sciences, 2008, 'Severe Space Weather Events—Understanding Societal and Economic Impacts Workshop Report', http://www.nap.edu/catalog.php?record_id=12507, accessed April 20, 2012.

NERC, 2010, 'High-Impact, Low-Frequency Event Risk to the North American Bulk Power System', June 2010, http://www.nerc.com/files/HILF.pdf, accessed April 20, 2012.

NGDC, 2007, "Major magnetic storms 1868-2006 according to the AA* criteria", http://www.ngdc.noaa.gov/stp/GEOMAG/aastar.shtml

Newcomb, Simon, 1859, Diaries of Simon Newcomb for August 25-28 available at the Library of Congress, Manuscript and Rare Books Division.

NOAA, 2001, ' Severe Space Weather Storms Affect GPS Meteorology Network', April 23, 2001, http://www.esrl.noaa.gov/gsd/media/hotitems/2001/01Apr23.html, accessed April 15, 2012.

NOAA, 2004: Intense space weather storms: October 19 – November 07, 2003. Service Assessment, 60pp. http://www.nws.noaa.gov/os/assessments/pdfs/SWstorms_assessment.pdf

NOAA Magazine, 2007, ' Researchers find global positioning system is significantly impacted by powerful solar radio burst', April 4, 2007, http://www.noaanews.noaa.gov/stories2007/s2831.htm, accessed April 15, 2012.

NPCC, 1989, 'Procedures for Solar Magnetic Disturbances Which Affect Electric Power Systems', C-15, https://www.npcc.org/Standards/Procedures/Forms/Public%20List.aspx, accessed April 20, 2012.

Oak Ridge Laboratory, 2010, "Electromagnetic Pulse: Effects on the U.S. Power Grid', October 2010, http://www.ferc.gov/industries/electric/indus-act/reliability/cybersecurity/ferc_executive_summary.pdf, accessed April 20, 2012.

Odenwald, S.F., 2010, 'Introduction to space storms and radiation, in Heliophysics: Space storms and Radiation; causes and effects, Carolus Schrijver and George Siscoe eds. Cambridge University press, pp.15-41.

Odenwald, S.F., J. Green, W. Taylor, 'Forecasting the Impact of an 1859-calibre Superstorm on Satellite Resources: Part I', Adv. Sp. Res., v. 38, pp. 280-297.

Odenwald, S. F., and J. L. Green, 2007, Forecasting the impact of an 1859-caliber superstorm on geosynchronous Earth-orbiting satellites: Transponder resources, Space Weather, 5, S06002, doi:10.1029/2006SW000262.

Odenwald, S.F., 2007, "Newspaper Reporting of Space Weather: The end of a Golden Era", Space Weather, Vol. 5, No. 11, November 2007.

Odenwald, S.F, and Greene, J. L., 2008, 'Bracing the Satellite Infrastructure for the next Solar Superstorm", Scientific American, August, 2007.

Odenwald, S. F., 1999, 'The 23rd Cycle- Learning to live with an stormy star', (Westfield Press, Boulder).

Olander, M., 2003, "Media use among young people" http://www.civicyouth.org/research/products/youth_index_2006.htm

Paxman, Lauren and Dan Hyde, 2011, 'HSBC systems crash affects millions across UK', Nov. 5, 2011, http://www.dailymail.co.uk/news/article-2057657/HSBC-customers-outraged-online-b Accessed April 15, 2012.

Penn, M.J. and Livingston, W., 2006, ' Long term evolution of sunspot magnetic fields' IAU 273.

Pesnell, W. D., 2008, 'Predictions of Solar Cycle 24', Solar Physics, v. 252, pp 209-220.

Prescott, G.B., 1860, "History, Theory and Practice of the Electric of the Telegraph", (Ticknor and Fields: Boston), p.326.

Pulkkinen, A Lindahl, S, Viljanen, A and Porjola, R, 2005, 'Geomagnetic storm of 29-31 October 2003: Geomagneticaly induced currents and their relation to problems in the Swedish high-voltage power transmission system', Space Weather Vol 3, S08C03 2005

Pulkkinen, A A, 2004, " October 29-31, 2003 geomagnetic storm: geomagnetically induced currents and their relation to problems in the Swedish high-voltage power transmission system", Eos Trans. AGU, 85(47), Fall Meet. Suppl., Abstract SA42A-07

Radio Netherlands, 2011 'Solar flare disrupts RNW short wave reception', 08.09.2011, http://www.rnw.nl/english/bulletin/solar-flare-disrupts-rnw-short-wave-reception. Acessed, April 10, 2012.

Reinard, A., Henthorn, J., Komm, R. and Hill, F., 2010, 'Evidence that temporal changes in solar subsurface helicity precede active region flaring', Ap.J. Letters, v. 710, L121-125. .

Shortwave America, 2012, 'M9-class Long-Duration Solar X-ray Flare', 01.23.2012, http://shortwaveamerica.blogspot.com/2012/01/m9-class-long-duration-solar-x-ray.html, accessed on April 12, 2012.

Silverman, Sam, 2014 'Auroral Annotations', http://nssdcftp.gsfc.nasa.gov/miscellaneous/aurora/cat_ancient_auroral_obs_666bce_195 1/silvnts.txt, last accessed April 15, 2014.

Song, W.B., 2010, 'An Analytical Model to Predict the Arrival Time of Interplanetary CMEs', Solar Physics, vol. 261, pp. 311-320.

SpaceDaily.com, 2003, 'Solar eruption likely cause of power outage in Sweden', http://www.spacedaily.com/2003/031031111934.of7v5sft.html, accessed April 14, 2014.

Space Weather Canada, 2011, 'Geomagnetic Effects on Communications Cables', http://www.spaceweather.gc.ca/se-cab-eng.php,. Accessed April 18, 2012.

Stassinopoulos, E.G., Brucker, G.J. ,Adolphsen, J.N. and Barth ,J., 1996, "Radiation-Induced Anomalies in Satellites", J. Spacecraft and Rockets, V. 33 ,PP. 877-882.

Stenquist, D., 1914, "The Magnetic Storm of September 25, 1909", Ph.D Dissertation, University of Stockholm.

Stevens, M., 2007, "Newspaper History", (Colliers Encyclopedia) http://www.nyu.edu/classes/stephens/Collier's%20page.htm]

Tezzaron Semiconductors, 2003, 'Soft Errors in Electronic Memory', http://www.tezzaron.com/about/papers/soft_errors_1_1_secure.pdf. Accessed, April 15, 2012.

The Economist, 2009, 'Cutting the cord', August 13, 2009, http://www.economist.com/node/14214847, accessed April 20, 2012.

Thomas, B.C., Jackman, C.H. and Melott, A.L., 2007, "Modeling atmospheric effects of the September 1859 solar flare" ,Geop. Res. Lett., v. 34, L06810, doi:10.1029/2006GL029174.

Thorpe, Lewis, 1974, The History of the Franks. London: Penguin Books. p. 30.

Townsend, L.W. (2003), Carrington flare of 1859 as a prototypical worst-case solar energetic particle event, IEEE Trans. Nucl. Sci. 50, 2307-2309.

Tribble, A., 2010, 'Energetic particles and technology', in 'Heliophysics: Space storms and radiation- causes and effects', Carolus Schrijver and George Siscoe, eds., (Cambridge University Press; United Kingdom), pp.381-399.

Vampola, A. L., 1987, "The aerospace environment of high altitudes and its implications for spacecraft charging and communications". J. Electrostatics ,v. 20, p. 21.

Waugh, R., 2012, 'Biggest solar storm since 2003 pummels atmosphere, forcing planes to divert from northern routes ', January 25, 2012, http://www.dailymail.co.uk/sciencetech/article-2091586/Solar-radiation-storm-Flights-diverted-Earths-atmosphere pummeled.html#ixzz1rjThzvhb, accessed April 15, 2012.

Westmidlands.com, 2014, 'Welcome to the Millennium',
        http://www.westmidlands.com/millennium/1900/1925-1949/1938.html

Wilkinson, D. 1994, "NOAA's Spacecraft anomaly data base and examples of solar activity affecting spacecraft", Jour. Spacecraft and Rockets, vol. 31, pp 160-165.

Wilkinson, D.C. , Allen, J. 1997, "Satellite Anomaly Data Base", ftp://ftp.ngdc.noaa.gov/STP/ANOMALIES/

Wik, M., Pirjola ,R., Lundstedt, H. ,Viljanen, A., Wintoft, P. And Pulkkinen, A., 2009, ' Space weather events in July 1982 and October 2003 and the effects of geomagnetically induced currents on Swedish technical systems', Annals of Geophysics, V. 27, pp 1775-1787.

Wrenn, G. L., Rodgers, D.J. and Ryden, K.A., 2002, "A solar cycle of spacecraft anomalies due to internal charging", Annales Geophysicae, v. 20, pp. 953-956.

Xinhuanet, 2005, 'Solar storm interrupts China's short-wave radio transmission', 01.21.2005, http://news.xinhuanet.com/english/2005-01/21/content_2491819.htm, Accessed on April 12, 2012.

Xihuanet, 2011, 'Solar flare affects shortwave radio communications in southern China', 02.16.2011, http://news.xinhuanet.com/english2010/china/2011-02/16/c_13733621.htm, accessed on April 12, 2012.

Zang, Z. 1985, in Korean auroral records of the period AD 1507-1747, Journal of the British Astronomical Association, vol. 95 No 5, pp. 205-210.

www.ingramcontent.com/pod-product-compliance
Lightning Source LLC
Chambersburg PA
CBHW050713180526
45159CB00003B/1013